BENCH PRACTICES FOR WATCH AND CLOCKMAKERS

By HENRY B. FRIED
C. M. W., C. M. C., F. B. H. I., F. N. A. W. C. C.

SECOND EDITION

Library of Congress Catalog Card Number 74-80841

ARLINGTON BOOK COMPANY, INC.
2706 Elsmore Street, Fairfax, Virginia, 22031, U. S. A.

This edition, published in 1997 by Arlington Book Company, Inc., 2706 Elsmore Street, Fairfax, Virginia 22031, USA is a republication of the work originally published in 1954 and the 2nd Edition has been revised and updated to include new methods and tools in 1974 by Columbia Communications, Inc., 370 Lexington Avenue, New York, N. Y. 10017.

Published for

TRANTIQUES

3303 Edgmont Avenue

Brookhaven, Pennsylvania 19015

Phone: (610) 874-1183 Fax: (610) 874-9217

ISBN 0-9656219-0-1

BENCH PRACTICES

FOR WATCH AND CLOCKMAKERS

By Henry B. Fried

In this book are the methods the author has used for over thirty years as a teacher, lecturer and writer of watch and clockmaking. Known for his excellent draftsmanship in addition to his skill, Mr. Fried - America's leading horologist and teacher - has used these skills to illustrate the methods of doing a job. In many cases, he illustrates and explains how one job can be done a number of ways, allowing the reader to select the method best suited to himself and his available equipment.

The hard-to-find instructions in all types of hairspring work are found here and make this a book in itself. The tricks in correcting bent hairsprings, selecting new ones and vibrating, collecting and studding them are made remarkably clear with illustrations not found in any other text. Jeweling of all types, lathework, the use of various lathe attachments, screwmaking, pivoting of wheels and pinions, regulator repairs and the many uses of the staking set are explained with these and others in over three hundred and fifty beautiful drawings and illustrations.

Repairs to wheels with missing teeth are covered in a separate chapter in fine detail showing how this can be accomplished in the manner done by craftsmen. Repairs to regulators, uses of the universal faceplate, pivoting, jeweling tools, slide rest as well as various hand tools make this book even more useful.

To those who restore antique watches, this book is especially valuable as it not only truthfully pictures the proper tools but explains the procedures carried out by masters in restoration, leaving nothing to the imagination nor taking it for granted that the reader has prior training. Thus to the student as well, this book can be used as a standard text. It usually offers more than one way of doing a job.

A special chapter on dial repairs includes a variety of methods of replacing porcelain dials, and formulaes for low melting solder.

Hard-to-find instruction on the many types of hollow pinions and wheels is also included. A detailed index makes for easy reference.

THE AUTHOR

Henry B. Fried, Certified Master Watchmaker and Certified Master Clockmaker is a third generation watchmaker who studied with New York's finest craftsmen. He has held positions as head watchmaker and trade shop operator before being licensed to organize and teach the first horology class for the N.Y. City Board of Education, a position he held for over thirty years. He is the author of numerous textbooks and articles on watch and clockmaking and holds patents in this field. Mr. Fried also is an excellent draftsman, having drawn virtually every illustration in this book.

Mr. Fried is the Technical Director of *The American Watchmakers Institute* and is the Horological Editor of *The Jewelers Circular-Keystone*. He also conducts The Answer Box for the National Association of Watch and Clock Collectors and is the accredited horological consultant-contributor to both *Random House* and *Merriam- Webster* dictionaries. His articles have appeared in almost every foreign trade journal. He lectures extensively and has appeared on radio and television often in connection with horology. He consults for the industry as well as governmental agencies.

Mr. Fried received the Outstanding Achievement Award from *The United Horological Association* and the "Man of the Year Award" from the jewelry industry *(WMDJAA)*. He is an Honorary member and past president of the *American Watchmakers Institute*. He has been president of local, state and national trade associations and his numerous citations and awards from various state and national organizations as well as those in foreign countries make him the most recognized horologist.

A collector of all forms of horolgia, Mr. Fried is a founder and Fellow of *The National Association of Watch and Clock Collectors* and a Fellow of the *British Horological Institute*.

Books by The Author

- The Watch Repair's Manual
- Bench Practices for Watch and Clockmakers
- The Watch Escapement
- Principles And Practices for Waterproofing Watches
- The Electric Watch Repair Manual
- The Calvacade of Time
- Escapement and Repair Section of the 400 *Day Repair Guide (Terwilliger)*
- The Packard collection of Complicated and Unusual Watches
- The Universal Watch Parts Catalog

CONTENTS

Book I
COMPLETE HAIRSPRING PRACTICES

Book II

Book III

PREFACE TO THE SECOND EDITION

When the author's first book, The Watch Repair's Manual was published, its readers were impressed with the method of presenting the elementary skills to the learner. Many requests were subsequently received for additional volumes on the more specialized operations in watch and clock repairing. A great many asked for instructions in all types of hairspring operations, jeweling, pivoting and escapement practices and other hard-to-find instructions in other specialized areas of watch and clockmaking.

The purpose of this book therefore, is to serve as an answer to these requests for profusely illustrated instructions in the modern techniques presented here. The illustrations in this book were drawn by the author and show the manipulative positions, tools, equipment and aid in a better understanding of the text.

This book should serve as a textbook to the student and apprentice. To the watch and clock repairer, this should serve as a guide to accepted and efficient trade practices. To the restorer of older and antique timepieces, the practices illustrated are those which will enable him to maintain the integrity of the prized item.

Some repairers prefer to use an elaborate array of equipment in performing a particular task. Other would prefer simpler methods. It is for this reason that more than one method of doing a job was given in the exposition of the material presented. Thus it will be possible for the repairer to choose the method best suited to his individual style of working.

This second edition has been updated to include new methods and tools. This volume also has been expanded to include the correction of the bent hairspring, often previously thought to be the *secret tricks* of a few. The inclusion of a comprehensive chapter on the replacing of wheel teeth should make this valuable to the restorer of old watches and clocks since replacement parts are unavailable and a desire to retain as much of the original timepiece is of prime importance.

Fresh Meadows, N.Y. 1974 HENRY B. FRIED

BOOK I

VIBRATING A HAIRSPRING

THE complete job of vibrating consists of the following separate and individual operations.

1. Selecting the hairspring.
2. Pinning it to the collet.
3. Truing the spring after it is colleted.
4. Attaching the colleted spring to the balance, timing or "vibrating" the spring.
5. Studding the spring and centering it. If it is a Breguet or overcoil spring, it must be raised, levelled, curved and centered.

CHAPTER I

HAIRSPRING ASSORTMENTS

Before going into the actual operations of hairspring work, some general knowledge of hairsprings should be had. When a hairspring

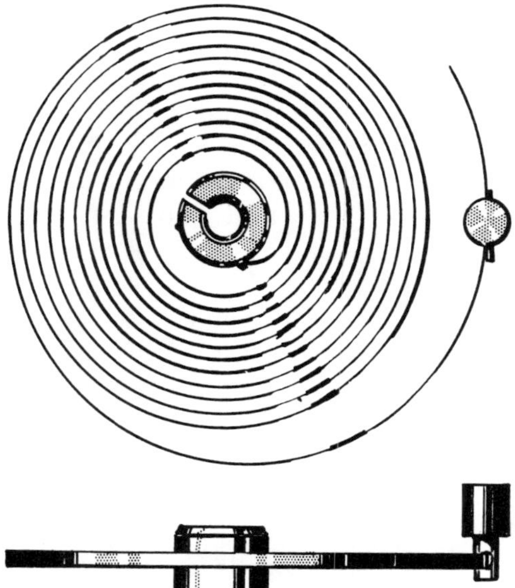

Fig. 1. Hairspring in which all of the spirals are on one level (flat hairspring).

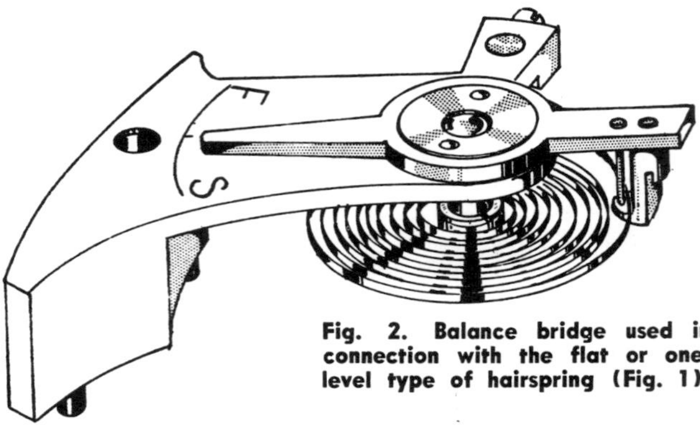

Fig. 2. Balance bridge used in connection with the flat or one-level type of hairspring (Fig. 1).

is to be chosen for a balance, there are many influencing factors. The diameter of the balance, the weight of the balance, the number of vibrations required in an hour, the type of hairspring (whether flat or overcoil) and the design of the balance bridge to which its outer terminal will be attached, all enter into the choice of spring.

When all the spirals of a hairspring are on one level it is called a flat hairspring as shown in Fig. 1. The style of the balance bridge which this fits is pictured in Fig. 2.

Fig. 3 shows a hairspring whose last coil is raised and brought over the main body. Such a spring is called a "Breguet," named after Abraham Louis Breguet who first introduced this form about 150 years ago in an effort to permit the spring to vibrate more freely and perform these vibrations in a concentric manner. The balance bridge of such a spring is illustrated in Fig. 4.

TYPES OF HAIRSPRINGS

It is possible to purchase hairspring assortments so that the selection of a spring is facilitated and the subsequent vibrating operation is more easily done. These sets are composed of a few dozen hairsprings, assorted according to the diameter of the balance. About twelve different sizes of balances for wrist watch sizes are represented on a gauge. "A" would be for the very smallest balance and going up to the largest balance in the assortment, let us say, "L."

However, balances of one size seldom equal each other in weight. One balance although of equal diameter may have a heavier or thicker rim, larger, heavier or more screws. Therefore, the heavier balance will offer greater resistance to the hairspring and the swings of the balance will be slower. To overcome this resistance, hairsprings in most assortments come in about four varying strengths for each balance size.

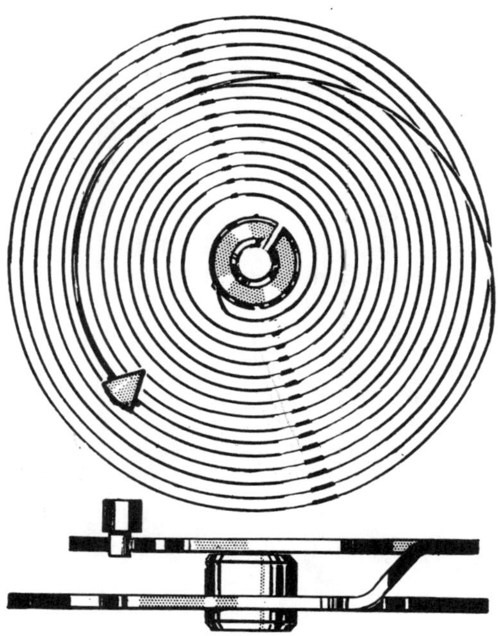

Fig. 3. Breguet hairspring in which the last coil is raised and brought over body.

Let us assume that we have a balance that fits into the gauge marked "H". If this balance came from a thin watch, it would also be thin and light. The hairspring most suited to this balance might be the one marked "H-1." A heavier balance of the same

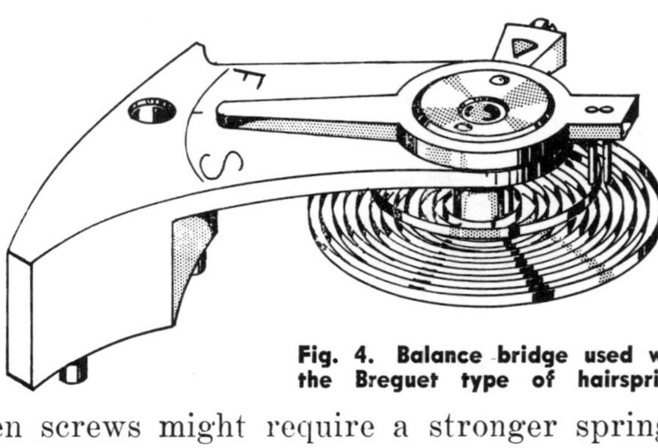

Fig. 4. Balance bridge used with the Breguet type of hairspring.

size with ten screws might require a stronger spring, let us say, "H-2." The same balance with sixteen screws might require the strongest spring, "H-4."

If springs from a mixed lot are to be used, there are a few rules which will help in selecting the proper spring.

Two main factors in choosing a spring are strength (in its relationship to the weight of the balance) and the diameter (in its relationship to the size of the balance and bridge.) Of the two, the strength of the spring must be given first attention. This is because a spring that first appears to be too large in diameter and with too many coils may also be too weak for the balance. Upon testing, however, it may be found that if a few coils were removed, the strength of the spring would be suitable. This would also reduce the diameter sufficiently to permit its proper attachment to the balance bridge.

Unfitted springs have between 14 and 18 coils. After fitting,

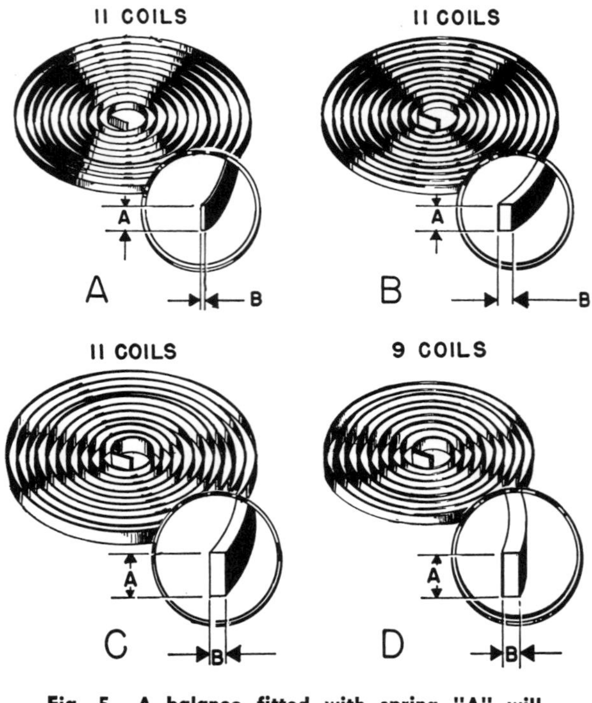

Fig. 5. A balance fitted with spring "A" will go slower than one fitted with spring "B." "C" is stronger than "B" because it is wider at "A." Spring "D" will act as the strongest because it is as wide and thick as "C." However, it is shorter and the arcs will be quicker.

these may finally have between 11 and 16 coils. This reduction affects its diameter as well as increases its resistance to the weight of the balance and increases the speed of the vibratory arcs.

The "strength" of a hairspring is dependent upon the number of coils to a given diameter (also called its length), the width and the thickness of the wire from which it is made. (Fig. 5).

If a hairspring were fitted to a balance and the end of this spring grasped with a tweezer and the balance permitted to hang, it would drop until the weight of the balance equalled the resistance of the spring. A lighter balance would not drop so far. A weaker hairspring would permit the balance to suspend itself further downward.

Balances that are required to vibrate a given number of times an hour (let us use the standard, 18,000) regardless of size, will have the same relative weight resistance to their hairsprings. This similarity in resistance of the balances to their hairsprings will manifest itself whenever the balances are suspended from their hairsprings.

Experiments or trials with a few balances of various sizes *already fitted with springs* (for 18,000 beat trains) will generally indicate the suspension distance to be between 5/16 to ½ inch as shown in Fig. 6. While this is true of balances requiring 5 beats a second or the 18,000 an hour, it is not a hard and fast rule. However, its observance helps in the initial selection of the spring.

To further understand the idea of the suspension distance being fairly equal for all size balances requiring the standard beat, an

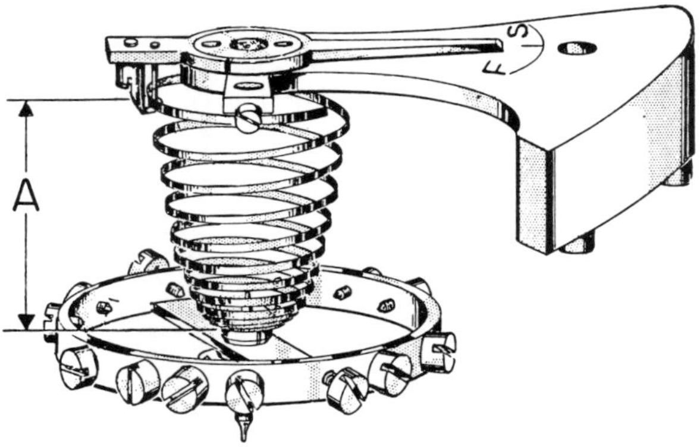

Fig. 6. Suspension distance is generally between 5/16 and ½ inch as shown at "A."

example will be given. Let us suppose that we fit a weak hairspring to a large heavy balance. Naturally, the balance will drop a considerable distance before it comes to rest—let us say about one inch. Realizing that the spring is too weak and the vibrations slow, a stronger spring is fitted. This will resist the weight of the balance and also provide faster vibrations. This is continued until a spring is fitted wherein the weight of the balance and the resistance of the spring are equalized, say at a suspended distance of about 3/8 inch.

Now let us try the opposite effect, placing a strong spring upon a small, light balance. This balance will hardly hang at all, only about 1/8 inch and its vibrations will be too fast. A weaker spring with the same number of coils is substituted. This allows the balance to hang lower and of course, the balance being comparatively heavier, now oscillates a little slower.

The lower a balance hangs suspended from the spring, the more it resists the spring and the slower it will vibrate. Therefore, balances that hang suspended from a hairspring a like distance from the point of suspension have something in common. That is, they resist the tension of their springs an equal amount and will vibrate at similar speeds.

THE SIZE OF A HAIRSPRING

So far, the strength of a hairspring has been illustrated. However, while a spring's strength may be suited for a balance, its size must permit it to be fitted to the balance and the balance bridge.

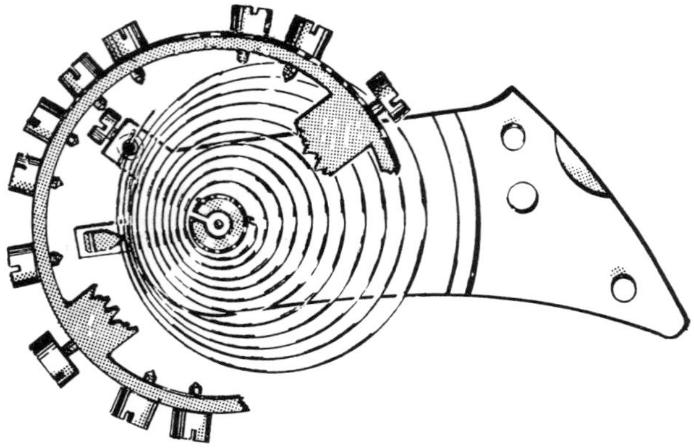

Fig. 7. A hairspring that is too large in diameter may not fit the balance bridge.

New uncolleted and unstudded hairsprings generally have about 18 coils. Hairsprings, factory fitted to balances and vibrated have an average of 12 to 15 coils. The uncolleted spring with its extra coils give it a greater latitude of use. To fit certain balances a few coils may have to be removed and the spring would still have close to the average number of coils.

The diameter of the average hairspring already fitted is slightly larger than half the diameter of the balance wheel. This means if a hairspring already vibrated was removed from the balance and

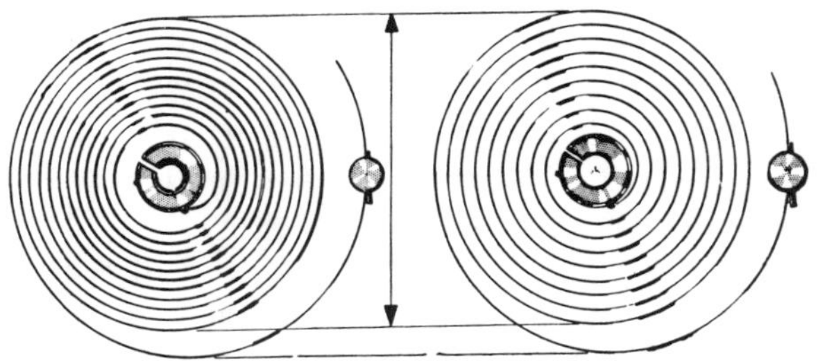

Fig. 8. Two hairsprings of equal diameter but with a difference in number of coils.

placed with one of its outer coils touching the balance staff, the extreme opposite part of that hairspring would just overlap the edge of the balance. Sometimes, an exception will be noticed in the case of a hairspring which exceeds these proportions. A hairspring that is too large in diameter may not fit the balance bridge. This is shown in Figure 7.

If the balance bridge is designed for an overcoil hairspring a larger spring may be used since it will not be as confined to a limited space. If it could be adapted it would have many advantages. A larger spring with many coils would not wind and unwind violently. This would prevent balance torque and aid in position timing. Also, regulation might more easily be obtained with the regulator since any movement of the regulator would not change the active length of such a spring to any appreciable extent.

Fig. 9. Many hairspring assortments are made without the collet attached in view of the wide variation in the size of collet holes.

On the other hand, hairsprings of comparatively large diameters have weight of their own and may "sag" in any position, creating errors. Large springs may also encroach upon the diameters of other parts and get caught or oiled. Oil on one coil would quickly transfer part of this liquid to a close neighboring coil. Also, some operators find them difficult to adjust since they require extra fine handling. Figure 8 shows two hairsprings of equal diameter but with a difference in number of coils. Notice the spacing of the coils in the two springs.

A short hairspring with few coils may be desirable because the coils are not close to one another and thus will not likely become stuck should oil catch one of the coils. It also would permit easier manipulation and be less susceptible to damage. On the other hand, a short hairspring is more violent in its unwinding and winding action. This might create a torque upon the balance staff and upset position adjustments as well as make adjustments for isochronism difficult. Also any movement of the regulator would create a great change in the general length of the spring and close regulation might be difficult.

In most cases, a medium is struck between the two extremes so that most balances have springs whose diameter is slightly more than half the balance and contain about 13 coils.

SELECTING A HAIRSPRING

The hairspring shoulders of balance staffs do not have a uniform thickness. Therefore, hairspring collet holes are not of uniform size. Because of this great variation in collet hole sizes many hairspring assortments do not come with attached collets. The

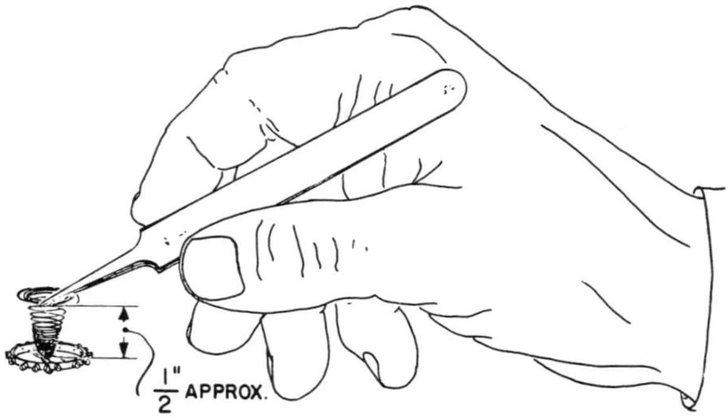

Fig. 10. If preliminary manipulations do not shorten diameter of spring to much less than half that of balance, the spring may be used.

uncolleted spring is shown in Figure 9. Since we know the required diameter and the general number of coils it needs, there still remains the problem of discovering the aproximate strength of the spring. We can test this by the method of suspending the balance from it and noting the height of the suspension.

To attach these uncolleted springs to the balance for a trial fit, a small bit of soft beeswax about the size of a pinhead is placed at the base of the balance staff. The inner coil of the hairspring is pressed into this temporary adhesive. This will hold the spring long enough to perform all preliminary observations.

Grasp the hairspring at a point one-half a coil or more from the outer end and permit the balance to hang so that the lower pivot barely touches the benchplate. Observe the distance from the point where the spring is grasped by the tweezer and to the point where the inner coil is attached to the balance. If this distance is more than one-half inch, grasp the spring closer to its center or further from the outer end, measuring the suspension and continuing to change the grasping point until this distance is just about one-half inch, as shown in illustration No. 10. If these manipulations do not shorten the spring so that its diameter becomes much less than half that of the balance, the spring may be used.

Before colleting the spring a final check on its size should be made. Turn the balance bridge over on its back and place the hairspring over it so that it is centered. This is shown in Fig. 11. Notice the coil that comes closest to the heavily dotted circle in this illustration. If the point where the hairspring was held last (preliminary vibrating point as shown in Fig. 10) is within this circle, the spring may be used.

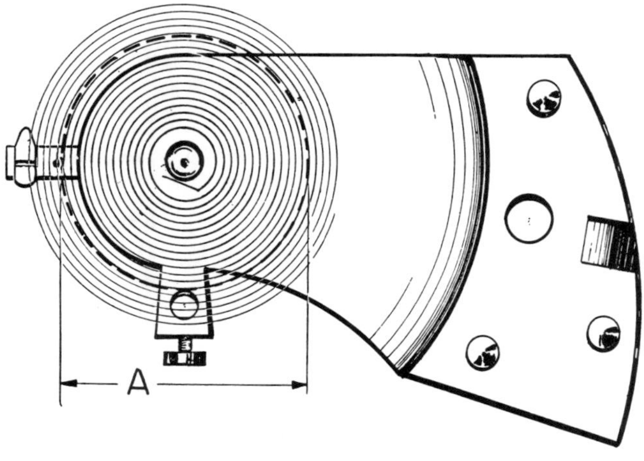

Fig. 11. Final check on spring size is made before spring is colleted.

Notice that in this illustration the heavy dotted line just touches the side of the regulator pin closest to the balance hole jewel. This diameter "A" indicates the largest size the hairspring may assume in order for it to fit the bridge properly.

Fig. 12 shows the spring colleted, studded and fitted to the bridge. Diameter "A" is the same as in Fig. 11, where the last coil touched the outside of the regulator pin. However, the last ¾ coil was bent outward to lie between the inside of the regulator pin and the (opened) regulator key. This last ¾ coil was re-curved to assume a circular, rather than a spiral shape. Now the full diameter of the spring becomes "B" in this illustration.

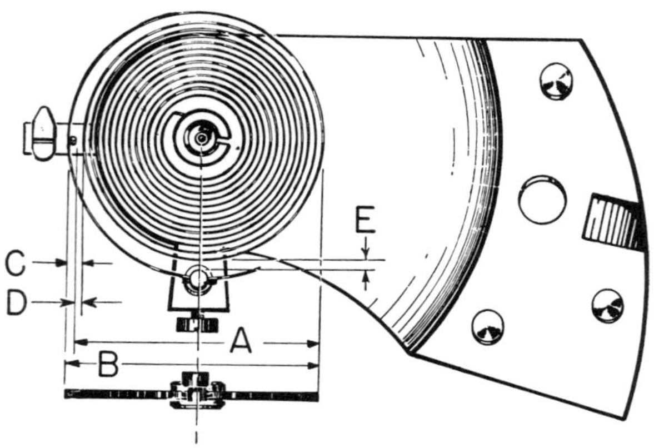

Fig. 12. Here, spring has been colleted, studded and fitted to the bridge.

In a fitted spring, it is a good rule to provide some space between the outside of the regulator pin and the second coil. Otherwise, the second coil might touch the regulator pin when the spring expands during the unwinding vibration. In this case, the diameter "A" indicates that the last coil touched the regulator pin prior to being fitted and re-shaped.

When the last coil was moved out to lie between the regulator key and pin, the space "D" was created. Instead of having the last two coils separated by space "D," now the spring has space "C." When this spring vibrates, there is the space "D" in which the second coil may move in and out. If it were any closer, it might touch the regulator pin and result in a very fast rate, especially when the mainspring is fully wound and the balance has a full motion.

Aside from the second coil touching the regulator pin, consideration must be given to the space between the next to the last coil and the stud. Space "E" provides some degree of safety at this point. "Flat" hairsprings, larger in proportion than here illustrated, will be crowded off center (see Fig. 7) and prevent any degree of accuracy in timing.

CHAPTER II

COLLETING

When it is decided to use a certain spring, it may have to be colleted. This entails the selection of a collet that will fit upon the staff properly, breaking out enough of the spring's inner coils to provide room for the collet, shaping this inner coil to permit it to enter the collet pin hole, pinning the spring to the collet and truing the spring both in the round and in the flat.

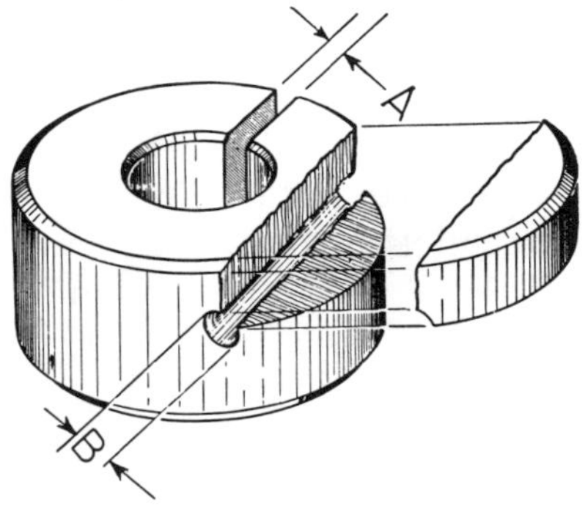

Fig. 13. Cutaway view of typical collet.

If the old hairspring is available, its collet should be used with the new spring because it has already fitted the balance staff properly. Fig. 13 shows a typical collet in a cut-away view. It is usually made of brass, split at "A" to provide a small degree of resiliency. Along its rim, a straight (lateral) hole "B" is drilled so that it is generally parallel with the slit in the collet. Into this hole, the inner coil of the hairspring is inserted.

A thin brass or nickel taper pin pushed into this hole alongside the inserted hairspring keeps the spring tightly in place.

TYPES OF COLLETS

Fig. 14 shows the two main types of collets used. 14A shows a collet used on a flat hairspring and 14B a collet generally used for a Breguet hairspring.

A B

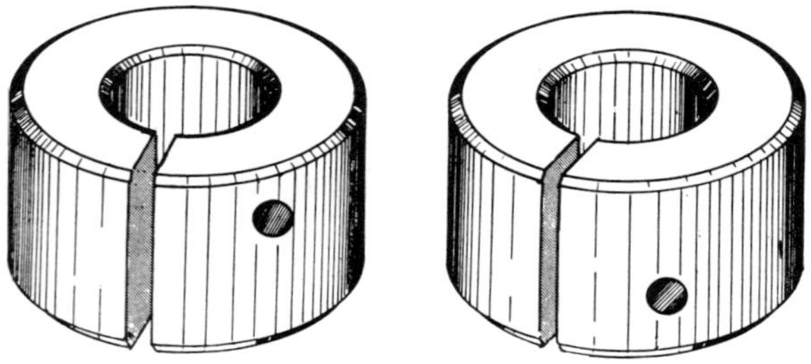

Fig. 14. Collet at left is for a flat hairspring; right is used with Breguet hairspring.

Collets for Breguet or overcoil springs have their lateral holes drilled below the center of the collet's rim. This is done to give more headroom for the two levels of this hairspring to operate between the confines of the balance wheel and the underside of the balance bridge.

Flat hairsprings do not need the height that Breguet springs require. Collets for flat springs therefore have these holes drilled along the center of the collet or above the center so that the spring will assume a position level with the hole in the stud (when it is fastened in the bridge) into which its outer coil is fastened.

If the old collet is to be used, the hairspring must be removed from it. This is accomplished by pushing out the taper pin holding the hairspring. To do this, simply, the collet is placed over a broach placed into a pinvise so that the point on the broach where the collet is fastened is just above the chuck of the pinvise. This is done so that greater control is assured as well as preventing the brittle broach from breaking due to pressure against the collet.

To remove the old collet pin, a sharp steel needle held in a pinvise or similar holder is pushed against the stump of the thinner end of the taper pin. This is most always adjacent to the end of the hairspring's inner terminal. (Fig. 15 shows how this operation is performed). The steel needle used should be thin enough to enter the collet sufficiently to eject the old pin. The point of this needle should be ground to an angle about 60°. If this point is made too

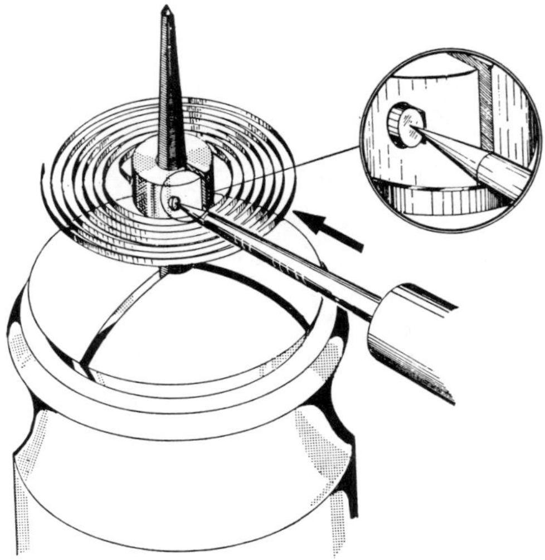

Fig. 15. Sharp needle, held in a pinvise or similar holder, is used to push out taper pin holding old spring in collet.

sharp, it might break or bend. The proper type of pusher-pin is shown in Fig. 15.

When the collet must be fitted to a new hairspring which has been selected by the methods previously described, some of the innermost coils must be broken out to provide space for the collet. This is because most new uncolleted springs have as many as 20 coils, with the innermost coils wound rather close to the center (see Fig. 9).

Breaking out some of these coils shortens the spring but since these coils are small diameters, their actual length is negligible compared to the length of an outside coil. However, too many coils should not be removed since the nearer we get to the outside, the larger the coils become.

To determine the correct amount of space needed for the collet, it is placed over the center of the hairspring as shown in Fig. 16. The exact spot where the innermost coil protrudes from beneath the collet is the point at which the hairspring is parted. This spot is indicated by the arrow in 16A. Generally the space between the outside of the collet and the first coil should be equal to a little more than the distance or space between a coil of the hairspring and its neighboring coil. If the hairspring is broken as directed, the tongue to be formed will shorten the spring sufficiently to provide the additional space required. After providing for the

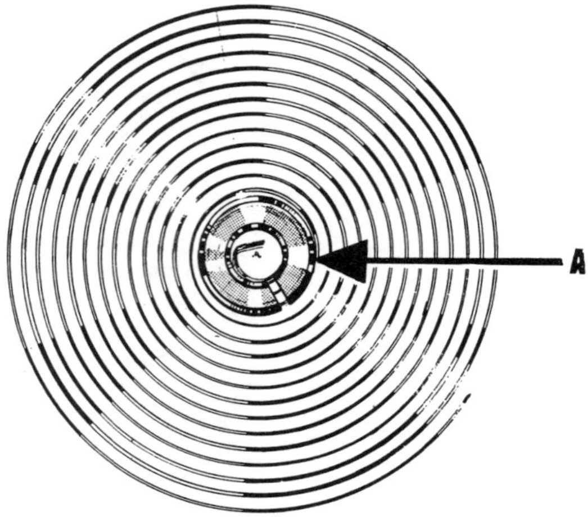

Fig. 16. Point where hairspring protrudes from beneath collet is place at which it is cut to allow space for collet.

collet space, the tip of the new inside coil is formed into a tongue. This tongue is shaped to enter the lateral hole in preparation for

Fig. 17. After hairspring is cut, end is shaped into tongue to fit into pinning hole.

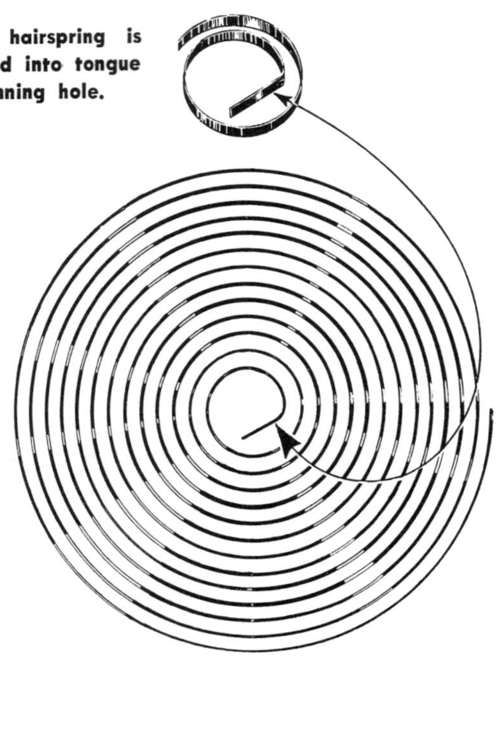

the pinning. This tongue is bent as shown in Fig. 17. For experienced hairspring operators with a trained eye, the spring may be colleted, trued in the flat and round with some degree of accuracy while the collet is held on a broach. To aid those who are not called upon to perform this job frequently, some aids and more elaborate devices are used. One of these is a colleting tool described in Fig. 18.

This tool is made from a length of brass bushing wire (A). A slight shoulder (B) is turned off to enable the seating of a seconds bit from an old 18s or 16s porcelain dial (C). This seconds dial is held to the main dial by bismuth, shellac or soft solder and it will require only mild heating to remove it from the larger dial.

To keep the dial in place upon the brass rod, shellac is melted around its base (D). Through the tubular hole a broach (E) is inserted. A set screw (F) holds the broach in place.

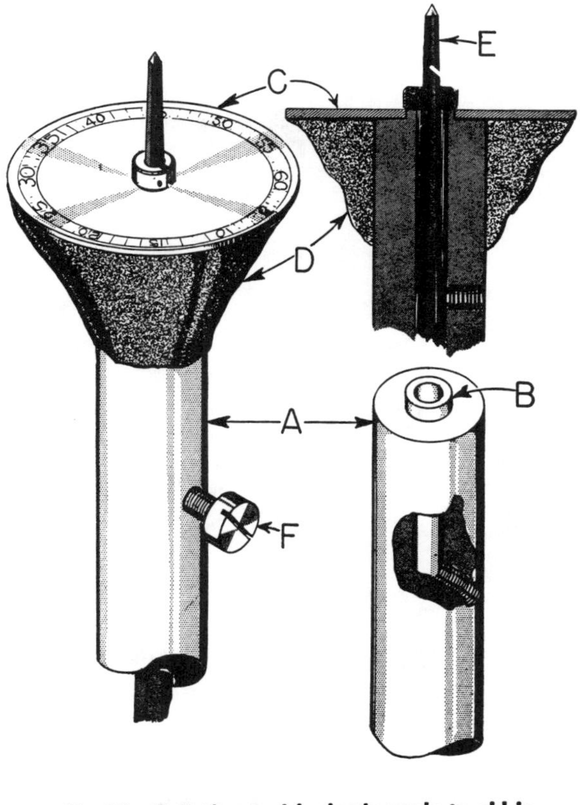

Fig. 18. Colleting tool is simply made to aid in colleting hairspring and truing it in the flat and round.

The collet is placed over the broach until it is tight. The broach is then lowered until the collet is close to the white dial's surface where it is fastened by the set screw. In this manner, the hairspring

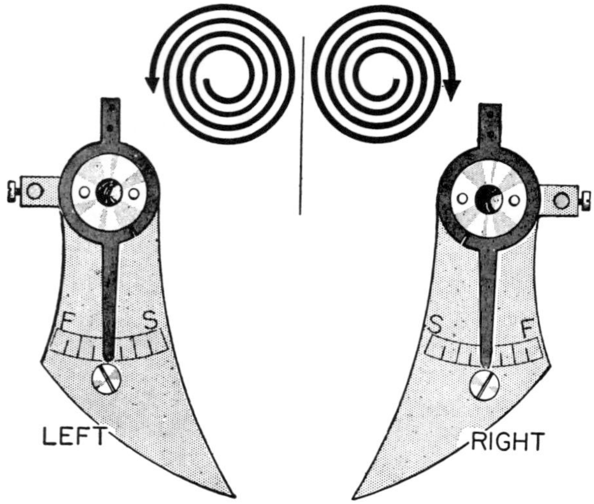

Fig. 19. Direction in which hairspring develops must be considered depending on balance bridge.

has a clear white background. Pinning and truing is then made easier.

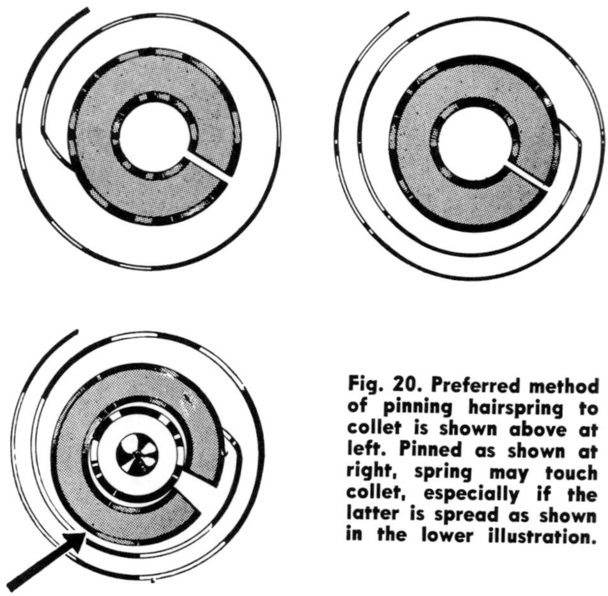

Fig. 20. Preferred method of pinning hairspring to collet is shown above at left. Pinned as shown at right, spring may touch collet, especially if the latter is spread as shown in the lower illustration.

Before placing the collet upon its holder, it must be remembered that most collets have their lateral holes drilled above or below their middle and the remarks previously made in regard to flat and over-coil hairsprings must be observed. Furthermore, the direction in which the hairspring develops must be considered. This depends upon the design of the balance bridge. Fig. 19 shows two balance bridges. One requires a *Right* hairspring and the other develops to the *Left.* Right hairsprings develop out of their collets in a clock-wise direction and *"left"* springs develop out of their collets in a counter-clockwise direction.

Should the collet's lateral hole be drilled midway between the top and bottom of the collet and it is parallel to the split, then some consideration should be given to the side of the collet from which the spring is to develop. Fig. 20 shows two springs pinned to the same collet. Upper left shows the preferred method. This will permit the spring to develop almost ¾ turn before it is opposite the split. In this way, the coil will spiral outward and be a considerable distance from the split in the collet. Furthermore, when the pin is pressed tight the pressure will not spread the collet apart.

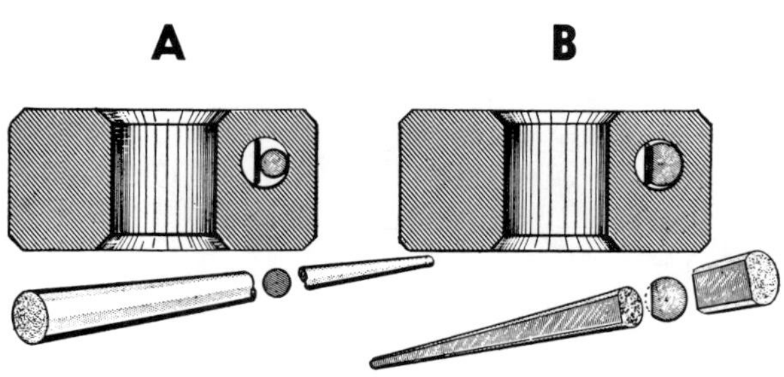

Fig. 21. Taper pin is filed flat on one side to provide a greater contact with hairspring than if it is left round as shown at the left.

When the spring is pinned so that it develops from the split end, as shown at right, there is a danger of its touching the collet especially if the collet is spread when being staked to the staff. This is shown in the lower left illustration.

HAIRSPRING PINS

Hairspring (collet and stud) pins may be purchased for this purpose and are very cheap in price. If the colleting job is to be done so that the spring is level and tight in the collet, a few preliminary operations are recommended. One of these is to make a trial fit with a short sample section of the hairspring to be fitted. This may come from the inner section previously broken out to provide room for the collet. In this way the taper pin can be prepared so that it will break off cleanly and exactly at the point desired when the final colleting operation is done.

To prepare the taper pin, it is first filed slightly flat on one side as shown in 21B. The advantages of providing a flat surface when this pin is inserted into the hole with the spring are considerable. One is that more of the pin enters the hole and greater surface contact is made both with the surface of the spring and with the hole in the collet. Compare 21A where the cross section of the collet and spring is shown with a regular round taper. In 21B, the same pin is filed slightly flat. Here a greater diameter of the pin is clearly shown, making its advantageous contact with the hairspring.

Furthermore by twisting this pin prior to being fully tightened and broken, the spring may easily be levelled.

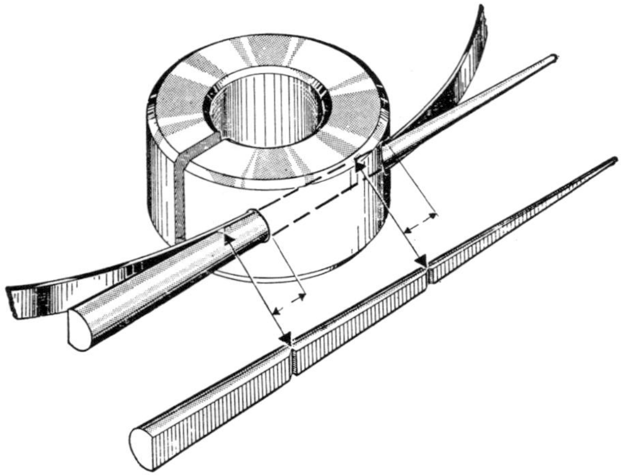

Fig. 22. V-grooves are filed in pin when it is fitted to facilitate breaking it off after the pin is finally inserted.

Place the trial section of spring in the collet and then place the pin into the hole so that the flat section is against the spring. Push the pin in as far as it will go without too much pressure. With a screw-head file cut a small "V" groove into the flat part of the taper pin a little above the spot where the pin enters the collet hole. The other groove is filed at the thin end of the pin indicated by the arrow in Fig. 22, after it is removed from the collet. The purpose of these "V" grooves is to facilitate the breaking off operation after the pin is finally inserted.

The spring is then inserted for the final colleting operation. The tweezer grasps the spring as shown in Fig. 23 and it is eased into the hole. The colleting tool may be turned to help meet the spring.

The taper pin is inserted into the hole beside the hairspring so

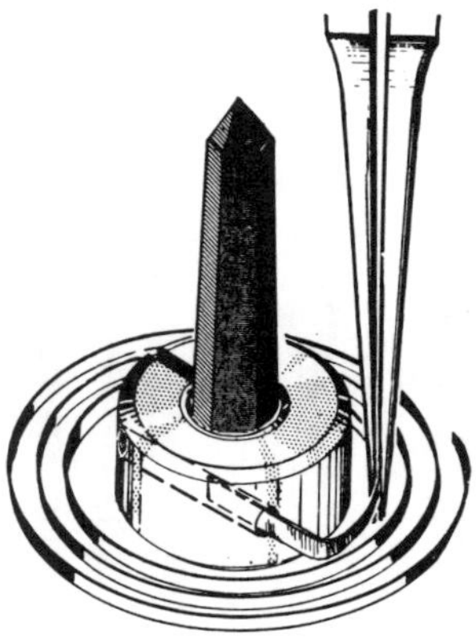

Fig. 23. Tweezer holds the spring close to its "tongue" and guides the spring into the collet hole.

that the flat side of the pin is against the spring. It is pushed in as far and as tight as it can be done with the tweezers. If the spring is tilted as shown in Fig. 24, twist the pin the direction of the arrow or in any direction which is necessary to level the spring. It

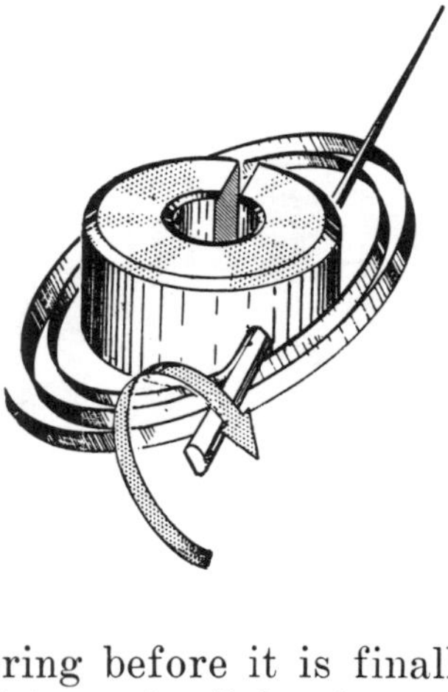

Fig. 24. If the spring is tilted as shown at left, it may be levelled by twisting the flattened pin in the direction of the arrow.

is best to level the spring before it is finally tightened because a spring bent or twisted to make it level sometimes wobbles during vibration in the watch. Furthermore, the hairspring must be inserted into the hole just the right distance. If the spring is inserted

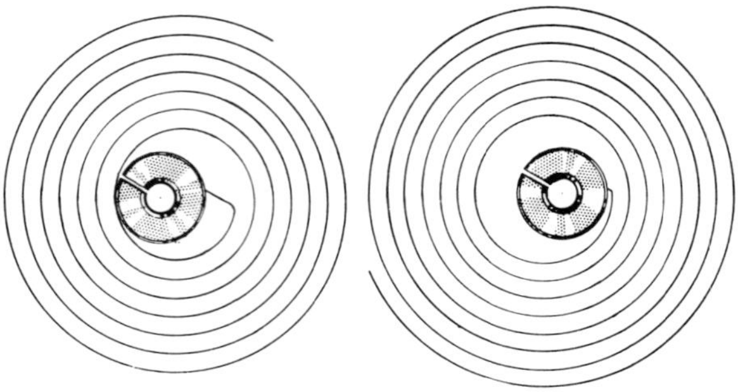

Fig. 25. Hairspring on the left was pinned out too far. Hairspring on right was pinned with tongue in too far.

too far or not far enough, it will be eccentric with the collet's center. Fig. 25 shows this clearly.

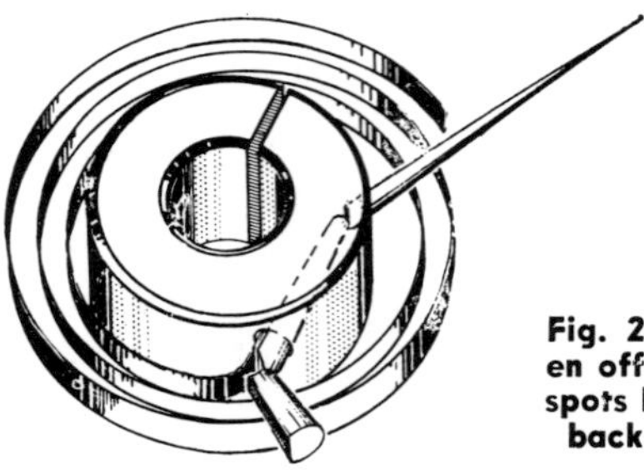

Fig. 26. The pin is broken off at the pre-grooved spots by bending the ends back as shown at left.

After all preliminary requirements have been met, the pin is broken off by bending it outward. It will break off at the places previously grooved as shown in Fig. 26. It will be noticed that the

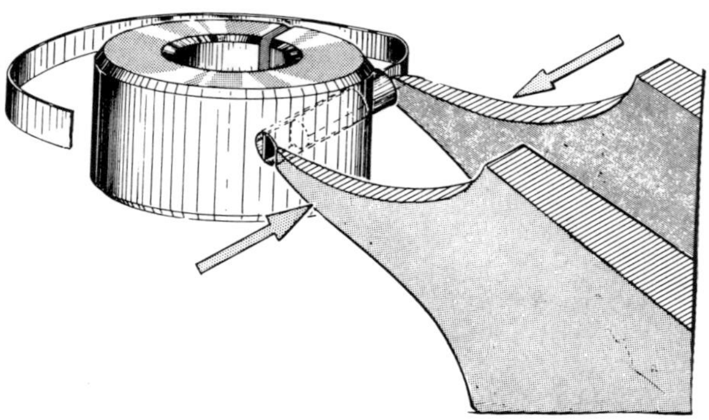

Fig. 27. A strong, stubby pair of tweezers squeezes the pin tight by bracing one tip on the thick stump of the pin and the other tweezer tip at the opposite hole and squeezing it together as shown above.

pin is broken off at its thick end at a point just outside the hairspring hole. At the thin end, this break should occur just *inside* the hole.

A stubby pointed tweezer is used to squeeze the pin in tightly. One point of the tweezer is placed against the thick end of the pin protruding from the entrance of the hairspring hole. The other tweezer point is braced against the collet, close to the opposite end of the hole, as shown in Fig. 27. Squeezing the tweezer blades together will push the pin inward until it is flush.

TRUING THE HAIRSPRING

PART I

AFTER the spring is fastened, it must be trued in the round and in the flat.

Hairsprings which have just been colleted may develop (spiral) out of the collet in a manner which indicates that the collet is not in the center of the spiralled spring.

This is due mostly to the central or innermost portion of the coil being pinned in too far or it may be too short or the inserted tongue may not be bent to the correct angle. It also may have been fastened

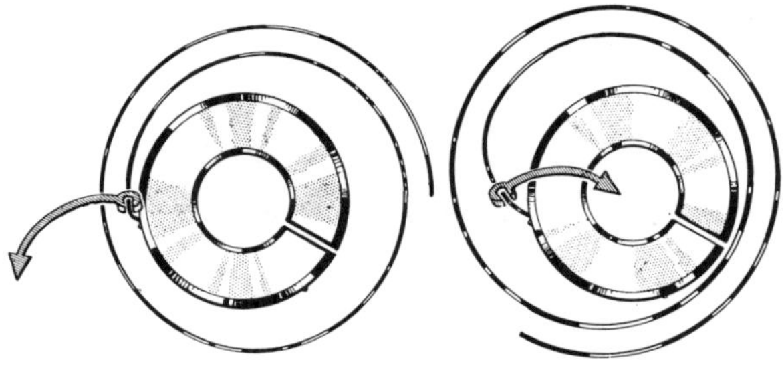

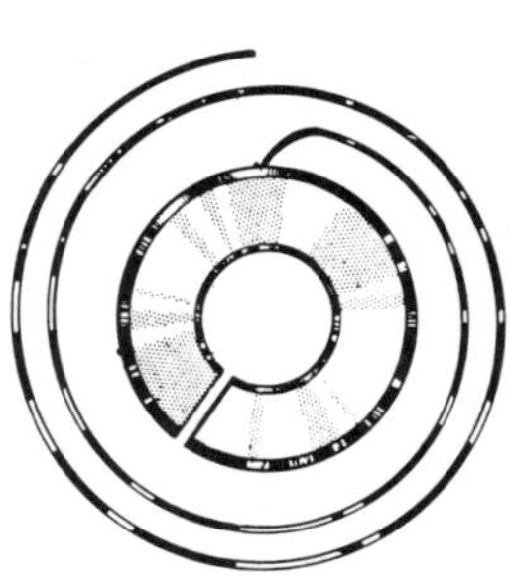

Fig. 28A. Above, left—Pinned in too far. To correct, tweezer twists spring out at the spot and in direction shown by arrow. Above—The first correcative bend makes spring look like this. Tweezer then bends in spring at spot and in direction shown by arrow. Left—Spring is brought back to normal via two bends shown at top.

while being tilted or there may be a combination of all these basic errors.

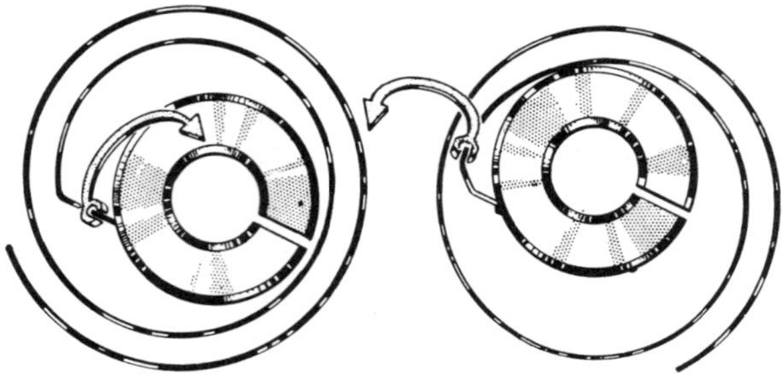

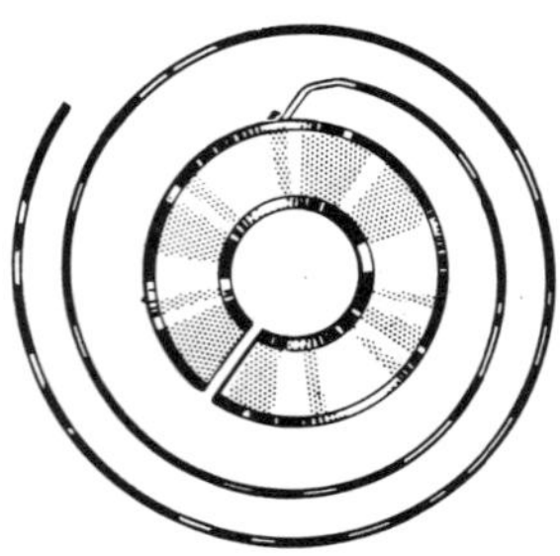

Fig. 28B. Above, left—Spring was not pinned in far enough. Tongue is too long. First corrective bend is made by tweezers at prongs of arrow twisting in direction shown. Above — First corrective bend is made as shown with tweezer pivoting at arrow prongs and twisting in direction of arrow. Left—Spring is now centered.

Since these faults fall into definite categories, some of these will be pictured with the suggested corrections in Fig. 28 A-B-C-D-etc. The colleted spring should then be placed on the balance.

USE OF HAIRSPRING TRUING CALIPERS

In order to determine accurately whether the hairspring is centered and is level, it is slowly spun between the jaws of a hairspring truing caliper.

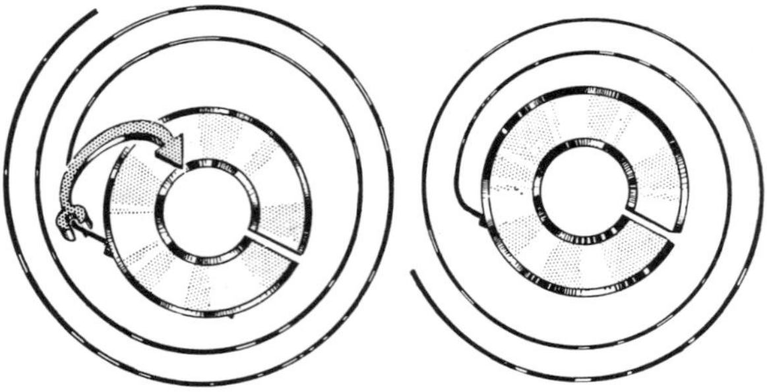

Fig. 28C. At left—Tongue is not bent enough. Tweezer grasps spring at prongs of arrows and twists in direction of arrow. At right—Spring corrected.

While the regular balance truing calipers may be used, their jaws are too massive to permit close observation of the innermost coils.

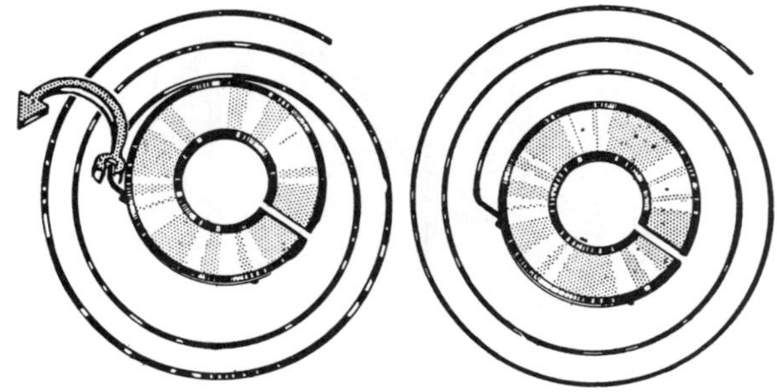

Fig. 28D. Spring is bent in too sharply. Place tweezers at prongs of arrow and twist in direction of the arrow.

An efficient truing caliper for this purpose may be made from old style frame calipers pictured in Fig. 29. To obtain a "bomb-sighter's view" of the inner coils, one of the jaws is cut away and is replaced with a strong but narrow strip of mainspring. The end

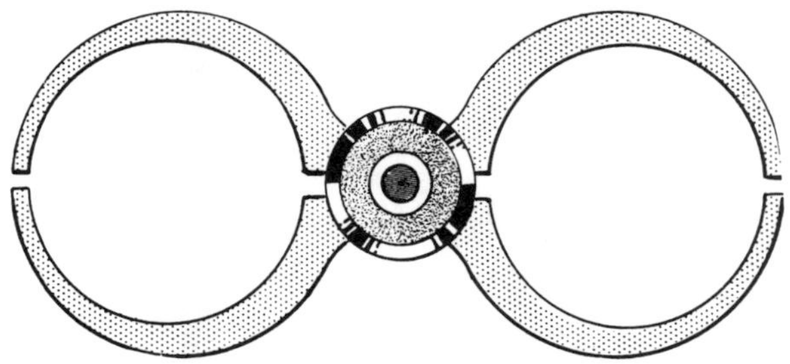

Fig. 29. Old-style frame calipers.

of this spring is pierced with a diamond pointed drill. The hole made by this drill will accommodate the upper pivot.

This strip of mainspring is either soldered or screwed to the frame as shown in Fig. 30C. This mainspring strip must be narrow enough so that the collet and the first few coils may be observed.

The balance and hairspring is placed in the caliper and slowly spun. With the eye loupe looking directly down into the spring, (Fig. 31) observe the first few inner coils while the balance turns

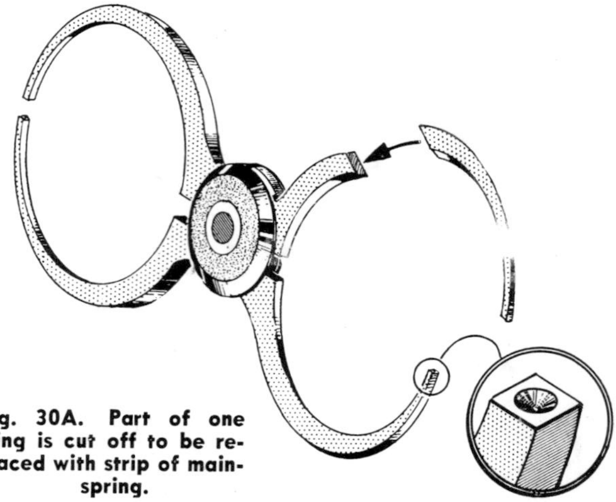

Fig. 30A. Part of one wing is cut off to be replaced with strip of mainspring.

slowly. If these few coils appear as concentric circles, it may be assumed that the spring is properly centered around the collet.

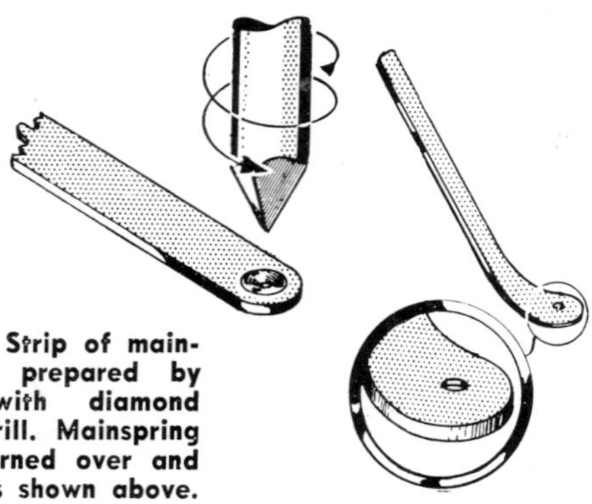

Fig. 30B. Strip of mainspring is prepared by drilling with diamond pointed drill. Mainspring is then turned over and finished as shown above.

On the other hand, if these coils appear to weave back and forth, in or out, it is an indication that they are not centered. Fig 28 should be then checked and the spring compared with it.

After these corrections are made, the caliper is turned so that the spring may be checked in the flat. Corrections made in the flat may disturb accuracies in the round and vice-versa. Therefore after every correction, the flat and round must be rechecked.

In a new spring, any deviation from a truly flat spring is generally due to inaccuracies at its pinning point in the collet. Corrections at the collet may be made either by lifting the spring at

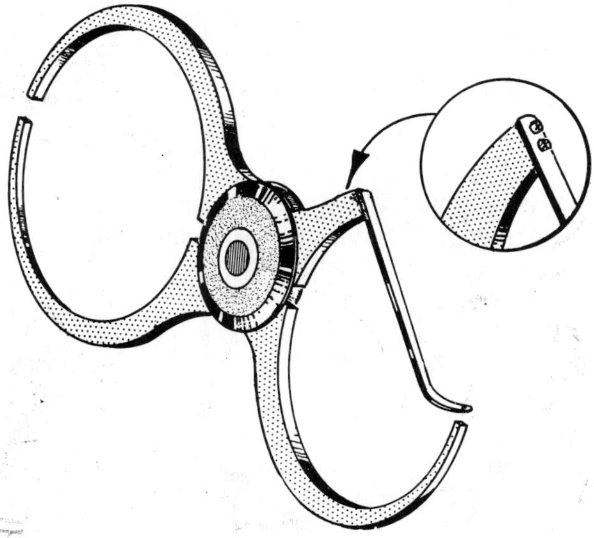

Fig. 30C. The strip of prepared spring is then either soldered or screwed to the wing of the caliper as shown.

the point where it emerges from the collet hole or by pushing it downward, depending on the correction required. This may be done with a needle or with a pair of fine pointed tweezers. Tweezers

Fig. 31. With eyeglass sighting the inner coils from directly above, observe behavior of spirals while balance is spun slowly.

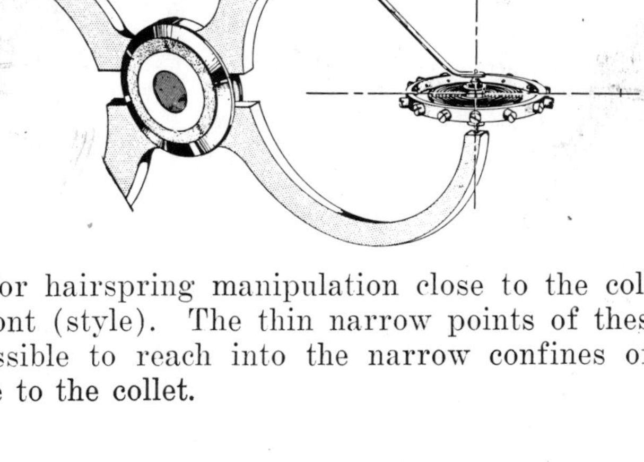

adaptable for hairspring manipulation close to the collet are the No. 5 Dumont (style). The thin narrow points of these tweezers make it possible to reach into the narrow confines of the hairspring close to the collet.

Chapter IV

CORRECTING THE BENT HAIRSPRING

Too often the young watchmaker is prone to abandon a bent hairspring. He sends the balance away to have a new hairspring vibrated or buys a complete new balance. This delays the job and decreases the profit.

It is better to use the old spring, since it was matched to the balance at the factory and you know that it has kept agreeable time.

With practice, one can become proficient at correcting bent hairsprings. The amazing facility with which women in watch factories and watchmaking teachers correct hairsprings is the result of constant practice.

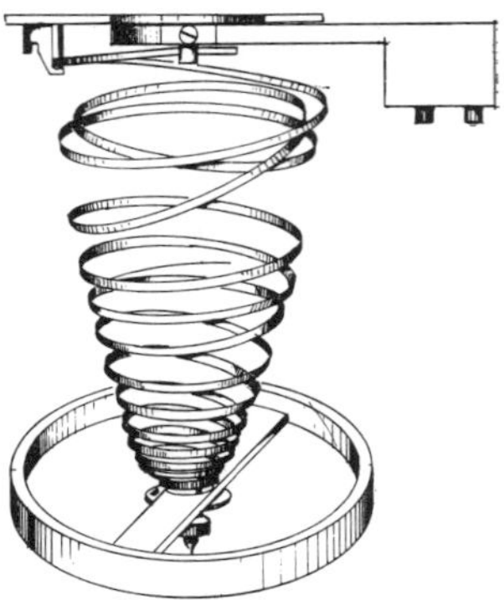

FIGURE 32: A "simple tangle" is actually not a tangle at all but an overlapping of one coil onto another. It is caused by a sudden jarring of the balance.

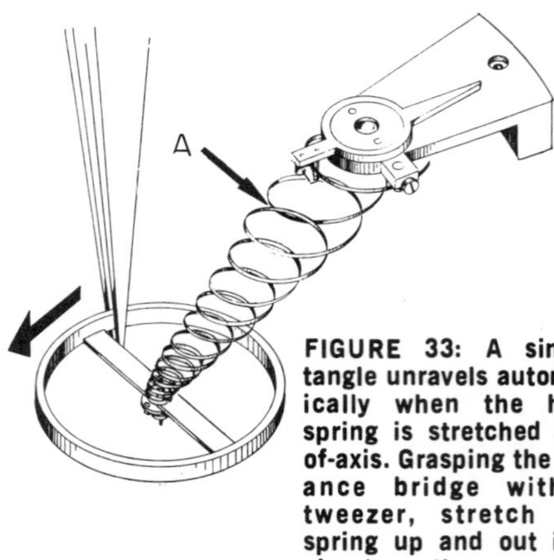

FIGURE 33: A simple tangle unravels automatically when the hairspring is stretched out-of-axis. Grasping the balance bridge with a tweezer, stretch the spring up and out in a circular path.

However, aside from practice, some logical procedures in correcting hairsprings will help.

Sometimes a hairspring has become tangled; one coil has overlapped another until it almost seems knotted. Most often this type of tangle is caused by a sudden sidejarring of the balance while it is out of the movement.

OVERLAPPED HAIRSPRINGS

A "simple" tangle is shown in Figure 32. If the hairspring and balance were as large as this figure, it would be easy to discover that this is not a tangle in the true sense but rather an overlapping of one coil onto another.

Untangling is simple. Grasp the balance bridge with the prongs of one tweezer and the balance with another or with the fingers of the other hand. Stretch the two apart considerably out-of-axis, as shown in Figure 33. The errant coil (designated by the arrow at *A*) will slip off the upper coil upon which it is resting and assume its natural position. Be careful not to over-flex the hairspring or it will become bent and distorted.

AVOID KINKING THE SPRING

If the tangle is situated low on the hairspring near the collet, it is difficult to free the errant coil without kinking the hairspring. Therefore, first bring the tangle closer to the stud or regulator.

Grasp the balance bridge with one tweezer and the balance rim with another or between the thumb and forefinger of the other hand. Pull the balance away from the bridge, causing the hairspring to become a descending, diminishing tapered helix. Now in this position, pull the balance to one side as in Figure 33, but move the balance in a circular path.

BRING NEARER THE STUD

If the hairspring develops from the stud toward the collet, clockwise, move the stud or its holder counterclockwise. This brings the twist nearer the stud and the larger coils where it can be more easily observed and manipulated.

If the hairspring develops from the stud to the collet, counterclockwise, turn the stud or its holder (the balance bridge) clockwise as in Figure 34. This brings the tangle nearer the larger coils at the top where it can be manipulated without bending or kinking the hairspring.

If the tangle develops close to the collet, the above process should not be used. Instead, the tangle should be eased up to a position closer to the outside of the hairspring.

USING A NEEDLE

First remove the balance from the bridge. Then use a fine, *thin* needle set into a pegwood holder, or your finest oiler ground to a polished point. Place the needle point between the coils, close to the collet. Gently follow a path outward in the space between the coils as shown in Figure 35.

When the needle comes to the point of the tangle, a gentle nudging outward will spread the tangle apart and ease it toward an outward spiral direction. Continue this until the tangle is at a position nearer the outer larger coils.

Now untangle the coils as explained previously or continue to work the needle outward until the tangle reaches the stud. The stud can then be manipulated out of the lower coils by holding the spring nearer the collet with one pair of tweezers and manipulating the closest coil with another.

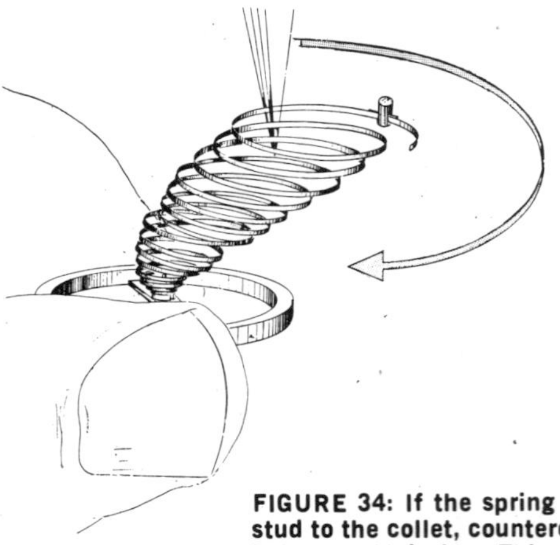

FIGURE 34: If the spring develops from the stud to the collet, counterclockwise, turn the balance clockwise. This brings the tangle nearer the larger coils at the top where it can be manipulated more easily.

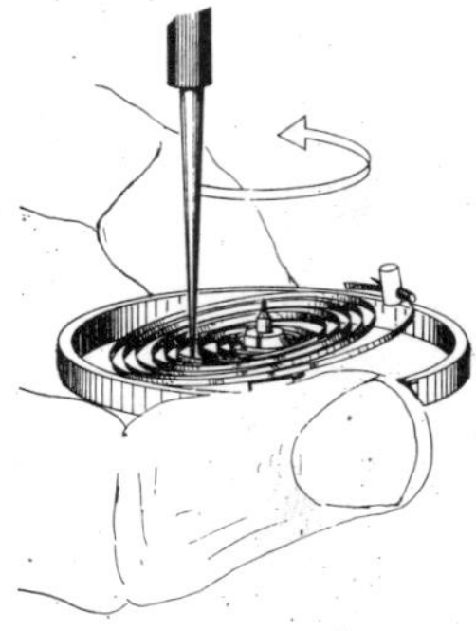

FIGURE 35: If the tangle is close to the collet, move it out to the larger coils by tracing a path with a needle. First, the balance must be removed from the bridge.

TANGLED HAIRSPRINGS

A true tangle is one which took place while the balance was disconnected from the bridge. The stud has threaded itself through the inner coils of the hairspring. Such a hairspring, if drawn tighter, becomes knotted. The tangle must be brought up to the stud and threaded through the last coil, using two pairs of tweezers.

A simple way of untangling a hairspring—preferred by many watchmakers—is to separate the tangled part from the untangled with a piece of white tissue paper.

Suppose the hairspring is tangled near its center, about the sixth coil. Allow the balance to hang from the hairspring, distending it. Insert the piece of tissue a coil or two below the tangle. Grasp the tissue paper in the fingers of one hand, allowing the balance to hang by the few coils below, supported by the tissue as a hanging shelf.

Now grasp the outer coils with a sturdy tweezer and twist in the direction which will cause the section of the hairspring above the tissue to thread itself below the tissue. This is shown in Figure 36. The tangle will be forced gently to the top outer coils near the stud where it can easily be unthreaded with another pair of tweezers.

CONSIDER THE COLLET

This method is not good when the tangle is near the collet. It may cause the closely-coiled inner spring to bend upward. In such a case it is better to use the thin needle, as illustrated in Figure 35.

If the spring contains more than one tangle or if the stud seems to be threaded into a few coils of the hairspring, it is wise to remove the hairspring from the balance. The temptation to force the hairspring coils apart to a distance greater than their natural resiliency might cause the stud to bend the hairspring.

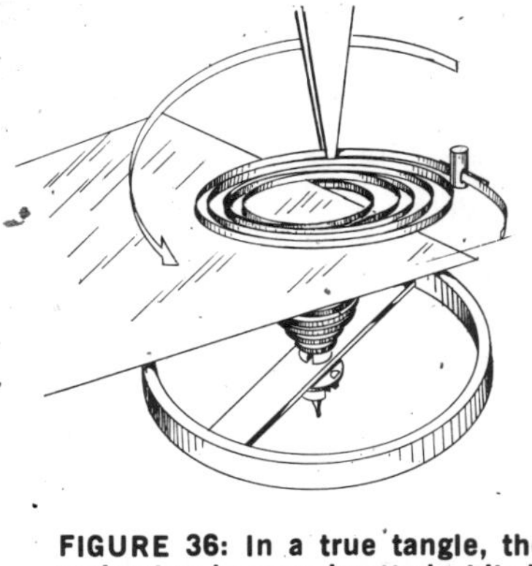

FIGURE 36: In a true tangle, the spring has become knotted while it was removed from the balance bridge. Use a piece of tissue paper as a shelf on which to hang the spring. Insert the paper a coil or two below the tangle.

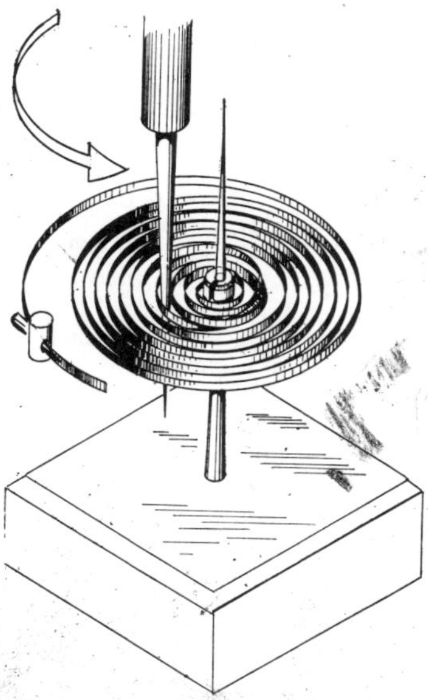

FIGURE 37: A good way to hold the hairspring when it is out of the balance is to mount it on a balance scaffold. Make one by setting a tapered pin into a heavy metal block.

USE A SCAFFOLD

A good way to hold the hairspring while out of the balance is to mount it on a balance scaffold, which you see in Figure 37. This is a tapered pin set into a heavy metal block. Place a white piece of paper over the pin to

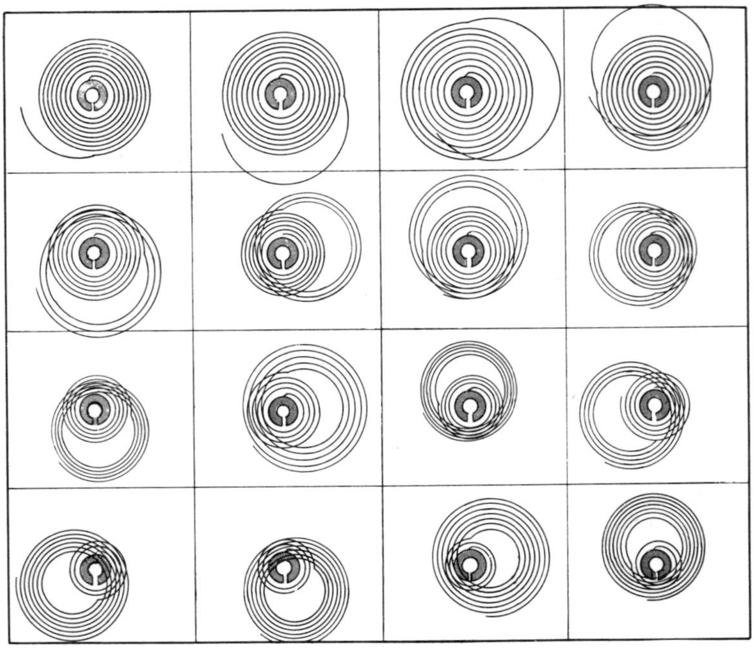

FIGURE 38

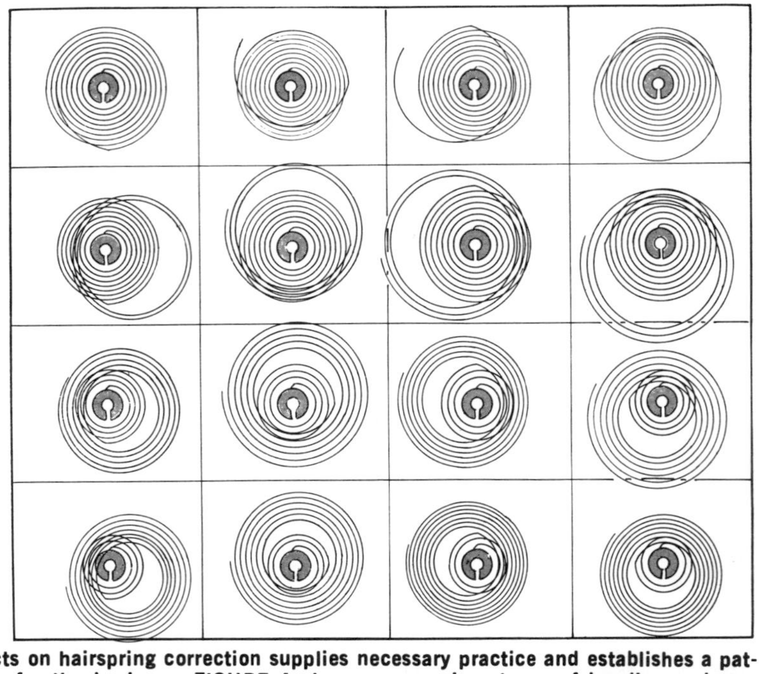

FIGURE 39

A series of projects on hairspring correction supplies necessary practice and establishes a pattern of procedure for the beginner. FIGURE A shows progressive stages of bending and correcting outward hairsprings. As the project brings the bend closer to the collet the job becomes increasingly difficult. After correcting a series of outward bends, the student is made to bend a spring inward and correct the bend one-quarter turn at a time approaching the collet. This is illustrated in the 16 diagrams shown immediately above in FIGURE 38.

provide a contrasting background which will reflect light. The scaffold allows the use of both hands, though one hand generally grasps its base. Be careful that the pin is not so long that it might injure the eye.

ONE TANGLE AT A TIME

Should the spring have more than one tangle, try to bring one of these tangles at a time up to the outside of the spring. Otherwise you will badly tangle the spring, kinking the coils. Sometimes when two coils are entwined closely, or two tangles contained in one, it is better to separate them singly. Even if these are outward near the stud, it is advisable to bring the inner tangled coil a bit closer to the collet. Then you will not be confused as you take care of the one outer tangle by itself.

NOTE STUD'S POSITION

If the tangle is so complex that it is confusing and the stud is in the way, remove the stud, first noting exactly its position on the hairspring for later replacement and correct timing. The tangling can then be unravelled without difficulty.

In some cases of tangling, the spring (when off the balance bridge) has become hooked by its stud around the rim of the balance wheel. This is easily recognized by noting whether a coil of the spring is looped over the circumference of the rim. In that case, grasp the stud in the tweezers and thread it out of the rim. If this happens in the old, split-compensation (bimetallic) balance, grasp the stud and ease the body of the spring near the stud back through the split in the balance.

KEEP UNIT INTACT

In practically all cases of tangling, it is better to allow the hairspring to remain on the balance. The hairspring can be handled more easily, and is not so sensitive to bending. Also, the weight of the balance assists in separating the coils in a vertical helix, making it easier to observe and select a particular coil with the tweezers.

Use the sturdiest tweezer that can discriminate between the coils. This permits a surer grasp of the spring and guards against kinking the coils. Make certain, however, that the tweezers are in good condition—smooth and without burrs—and that the blades of the tweezer are of equal length, width and thickness.

Chapter IV

BENT HAIRSPRINGS

Bent hairsprings are of two types: those which tilt from the main body and level of coils to occupy different levels, called "out-of-flat"; and those which cause the spiral development to become disorderly and unequal, called "out-of-round."

CHOOSING THE RIGHT TOOLS

Hairspring tools consist of finely pointed tweezers and long, thin, polished needles set into light holders such as pegwood or plastic sticks. Only the most delicate pointed tweezers should be used for work near the collet.

Experienced watchmakers and hairspring workers prefer short, pointed tweezers over the long tapered types. Short tweezers are stronger, yet can reach in between most coils of the hairspring. The strongest tweezers possible should be used, as they permit you to get closer to the work and they relay greater sensitivity while making corrections.

FIND THE BENDING POINT

The first step in attempting to correct a bent hairspring is to find the source of error, or, the bending point. The exact spot of the bend can be elusive because it might be hidden under another coil. In some cases the bend is so gradual that its exact beginning is difficult to find.

Figure 40 shows a hairspring bent outward in-the-round. The bend is a sudden, sharp deviation and easy to spot. Notice that when a spring is bent outward there is an area where the space between the coils is greatest. The source of the bend is one-quarter turn inward toward the collet.

On Figure 40 a cross-line illustrates a system for locating the deviation point. The area of the greatest space between the coils is at 90 degrees to the bend itself.

CORRECTING THE OUTWARD BEND

After determining the point of bending, grasp the spring with tweezers at this point but a little behind the bend, closer to the collet as shown in Figure 41. If the bend is near the outer end of the hairspring, nudge the hairspring with a finger of your free hand. You will have a more direct and sensitive touch than if the "feel" has to go first through a tool and then to the fingers.

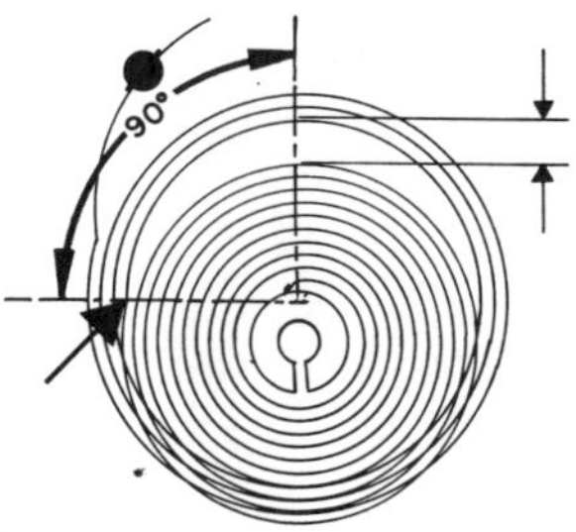

FIGURE 40

Before attempting to repair a bent hairspring, locate the exact bending point of the coil. You will notice with an outward bend that there is an area of greatest space between the coils. Progressing inward toward the collet, the source of the turn is 90 degrees, or one-quarter turn around, designated by the arrow.

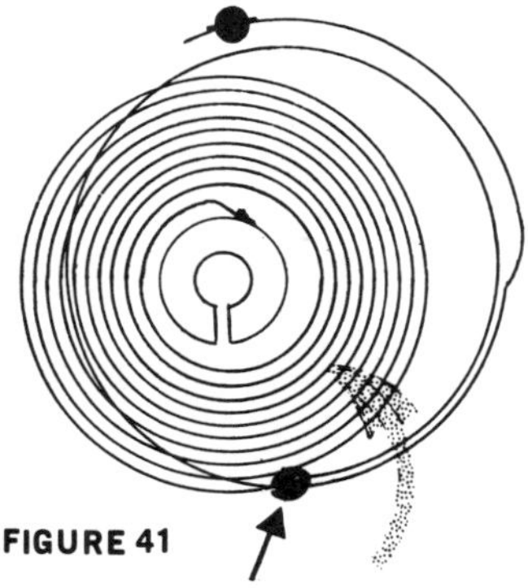

FIGURE 41

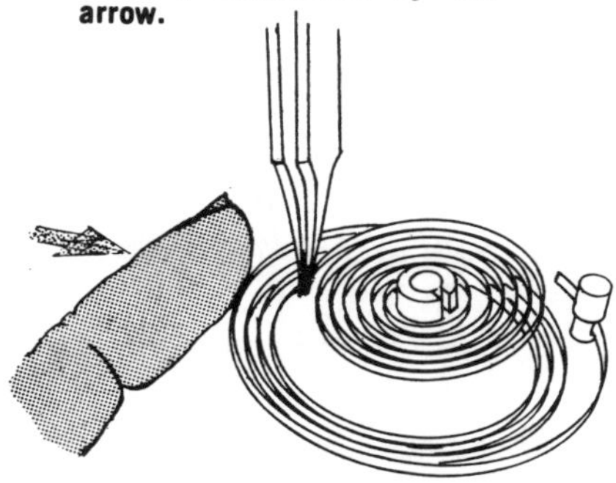

FIGURE 42

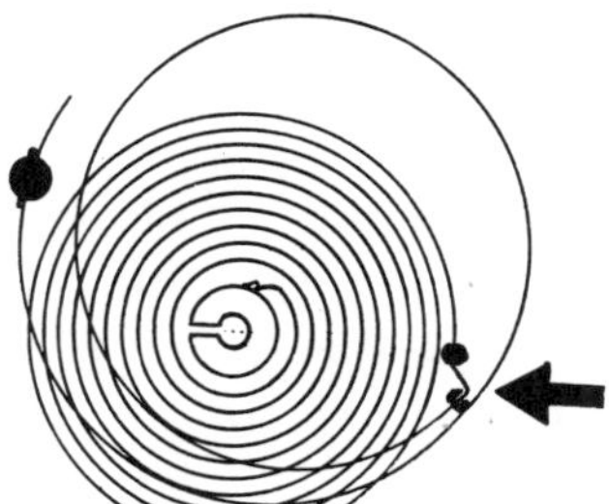

FIGURE 43

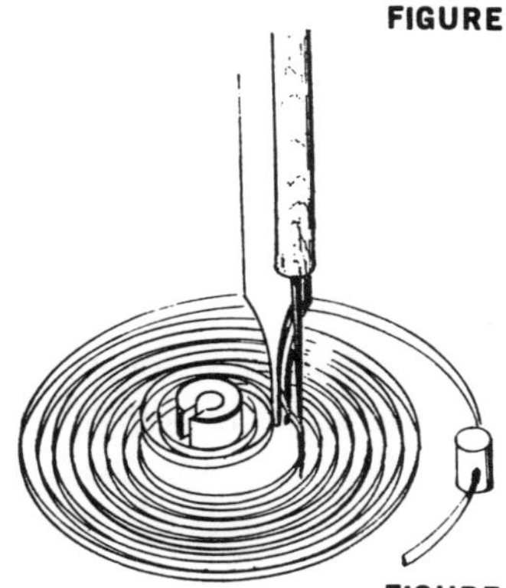

FIGURE 44

To correct an outward bend, grasp the spring with tweezers just behind the bend as shown in FIGURE 41. Using your finger, as in FIGURE 42, nudge the section of spring outside the bend. The tweezer acts as a vise. If you prefer to use a second pair of tweezers, use it closed. If both pairs grasp the spring the slightest shake will result in a new bend, as in FIGURE 43. A polished needle is best to nudge a bend so close to the collect as in FIGURE 44.

The tweezers acts as a vise, while you nudge the entire section of spring outward of the bend. Do not be concerned that the nudging finger is actually pressing on the outer coils, as in Figure 42. The spring will only bend at the point held by the tweezers. Using the finger whenever possible is the safest method. There is less chance of introducing a new bend in the "fault."

If you use two tweezers, grasp the hairspring with one and use the second closed, as you would a needle. If both tweezers are grasping the spring, any shake or miscalculation will result in a new bend, similar to that in Figure 43.

Rather than chance excessive bending or new bends in the opposite direction, nudge the spring lightly and repeatedly so that the bend back to normal is made in gradual increments.

COMPOUNDING THE PROBLEM

If the bend is close to the collet, do not use the finger to nudge the hairspring directly. A fine polished needle set into a pegwood stick is best. Grasp the hairspring with the tweezer, as usual. Sometimes if the bend is close to the collet it is advisable to grasp the hairspring on the side of the bend closest to the *stud* and not the collet. Then the needle can nudge the spring and collet back to normal gradually, as in Figure 44.

If there are two or more bends in the hairspring, correct the more severe bend first. This restores at least a partial appearance of an orderly spring and lessens the confusion of coils overlapping each other. If multiple bends cause confusion, the simplest procedure is to correct the outermost bends first, working gradually toward the center. Then more critical and exact corrections can be made afterward.

Again, a reminder—moderately pointed tweezers are recommended for correcting coils near the stud, while finer tweezers are needed closer to the collet. Near the collet the radii of the curves are shorter and sharper. Therefore, any corrective force exerted near the collet has a greater effect and causes a sharper bend than the same force applied to a coil near the stud.

When a coil has been bent inward, as in Figure 45, it looks in worse condition than one with an outward bend. This is because the coils cross one another and, as does Op art, confuse the eye as it follows a spiral path along the spring to pinpoint the bend. Actually, an inward bend is easier to correct.

Determine the bending point with the same 90-degree method described earlier. The one exception with an inward bend is that because the fault has caused a diminishing of space, we trace it from the point of *least* space back one quarter turn toward the collet.

Manipulate a pair of tweezers with flat inside surfaces so that the blades straddle the bend, as in Figure 46. Pinch the tweezers gently to bring out the bend and restore the curve. If you squeeze too tightly, the bend will be overcompensated. It will then have to be corrected as if it was an outward bend, which we described previously.

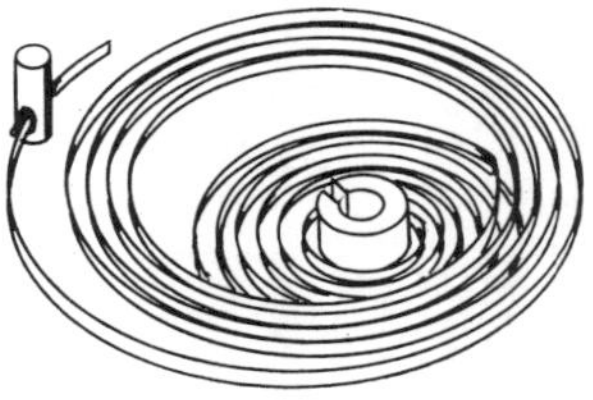

FIGURE 45

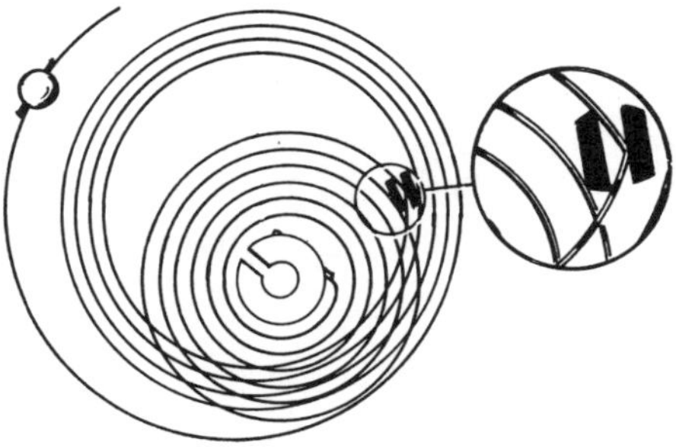

FIGURE 46

An inward bend, such as in FIGURE 45, looks more confusing because the coils appear to cross each other. Correct the bend by pinching gently between a pair of flat-surfaced tweezers, as in FIGURE 46.

THE GRADUAL BEND

If a spring has been bent so slightly that the bend is not sharp and cannot be spotted easily, it becomes more important to discover the exact point of deviation. Otherwise, an inexact point of correction will only introduce a new bend.

Follow the path between the coils outward from the collet with your eye until the space between the coils either diminishes or becomes greater. Then use the 90-degree method to determine the bending point.

Figure 47 shows an example of a bend which confuses efforts to correct it because it appears that there are a number of bends. Actually, there is only one bend, situated at *A*. This bend has caused the spring, one-quarter turn outward, to press against the adjacent coils and force them to touch one another at *B*. This, in turn, has caused the spaces diametrically opposite to become a like amount greater, at *C, D, E* and *F*.

In such a case, pinch the hairspring at *A* to bring the coil outward. This relieves the pressure against the two inner coils at *B* and diminishes space *F* back to normal.

Again, pinching the spring when a bend is closest to the collet has a greater effect than the same manipulation closer to the stud. A glance at Figure 48 makes this clear.

Gradual bends can be corrected by a series of small bends, each close together to form a gradual curve resembling the correct spiral. Sometimes the correction can be made with the finger, as in Figure 49. Diametrically opposite the anchored bend, push the hairspring with the finger to hump up that portion of the spring momentarily and restore the correct curve. Of course, take care not to overbulge the spring.

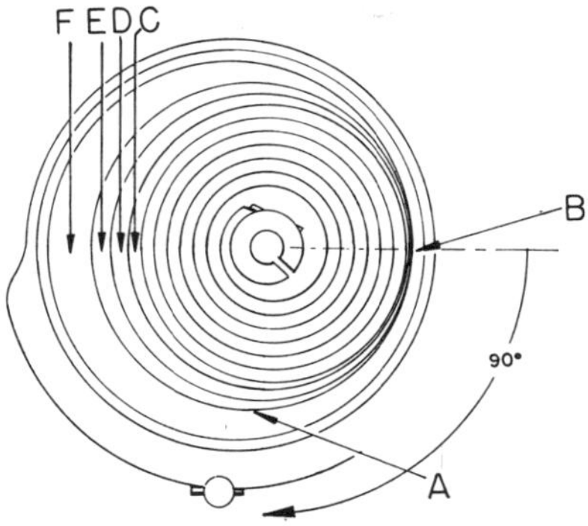

FIGURE 47

A gradual bend, at A, forces adjacent coils to touch one another at B, creating larger spaces at C, D, E and F.

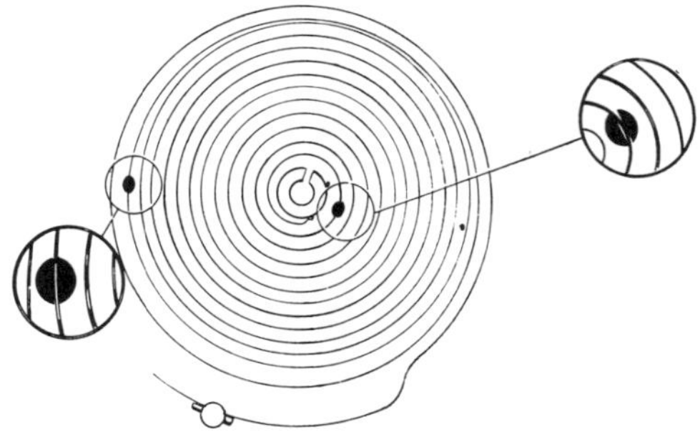

FIGURE 48

Pinching a bend in a coil near to the collet has a greater effect because of the more pronounced curve of the coil.

CRINKLED HAIRSPRINGS

Fortunately, crinkled hairspring sections are usually confined to the area near the stud. They are often caused by pinching from the pins of a nearby regulator. Inexperienced corrections further the crinkle while the spring is still attached to the balance bridge.

The corrective procedure is literally an ironing-out of the spring using two pairs of tweezers. With the stronger of the two, grasp the good por-

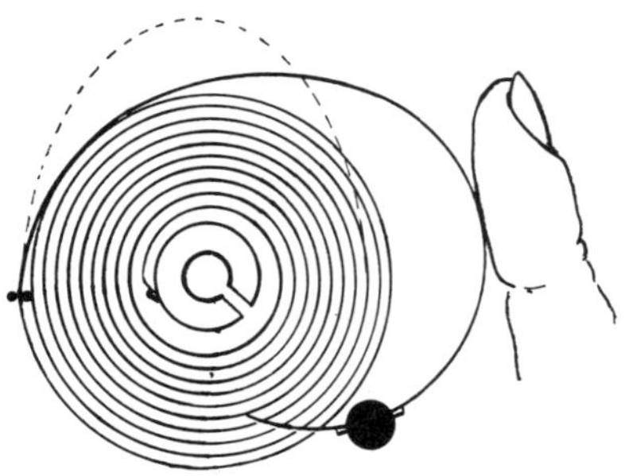

FIGURE 49

A gradual bend can be corrected by a series of small bends formed by humping the spring with the finger.

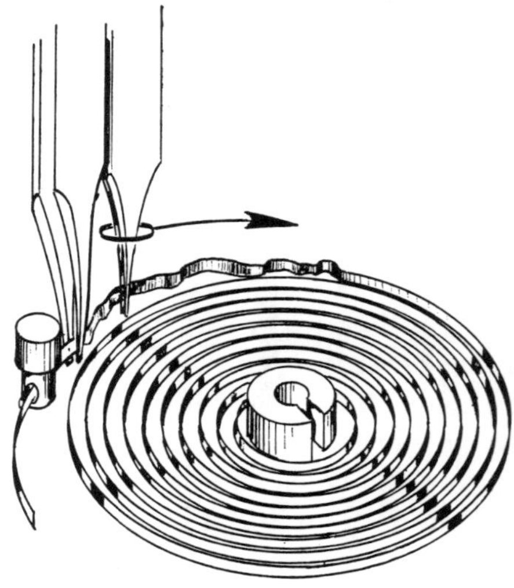

FIGURE 50

A crinkled hairspring is corrected by stretching or ironing out the coil between two pairs of tweezers.

tion of the spring tightly. With the weaker tweezer, preferably one with an oiled smooth inner suface, grasp the spring as close to the strong tweezer as possible and lightly move it away from the strong tweezer, as in Figure 50. Repeat this operation, increasing pressure on the spring until it is free from crinkles. Then correct the curve as with a gradual bend.

Some watchmakers correct crinkles by resting the spring edgewise on a polished piece of white glass. The glass provides a smooth surface and contrasting background. Then rub the hairspring with the back of a rounded tweezer or polished needle. While the needle irons out the crinkles, it also causes the hairspring to curl upward, assuming nearly the curve which it formed originally.

By all means do not grasp the stud with the tweezer while ironing out the spring with another tweezer. This will only result in tearing the spring at the stud.

If correction must be made close to the stud, grasp the spring so that at least a tweezer's point breadth of light can be seen between the tweezer point and the stud. Then turn the spring so the stud points downward, enabling more of the spring to be seen and pinched. Hairsprings are weakest at the points where they enter the collet and the stud because the pinning and bending operations have already put these points under stress.

Many studs have two diameters. The larger is on top, while the smaller one bearing the spring is below. Turning the spring on its face allows the tweezer to get directly at the point where the spring enters the stud.

Correcting hairsprings which are bent out of flat is more difficult than correcting those which are bent out of round. If you place an out-of-round hairspring on a flat white background, you can watch the fault while you correct it. But in order to see the fault in an out-of-flat hairspring you must lift the spring and view it from the edge, select the bending point from the other coils, then lay id down again and remember that exact spot while proceeding to correct it.

FIND THE BENDING POINT

First we must establish certain principles of procedure. When correcting in the round, the important factor was establishing the point at which the spring was bent. You will remember that this spot was generally found to be 90 degrees inward from the widest (or narrowest) spot between the coils.

When truing in the flat, the point at which the error is situated in the majority of bends is 180 degrees inward from the point of highest deviation between the two levels of the spring. Figure 52 appears as two sepa-

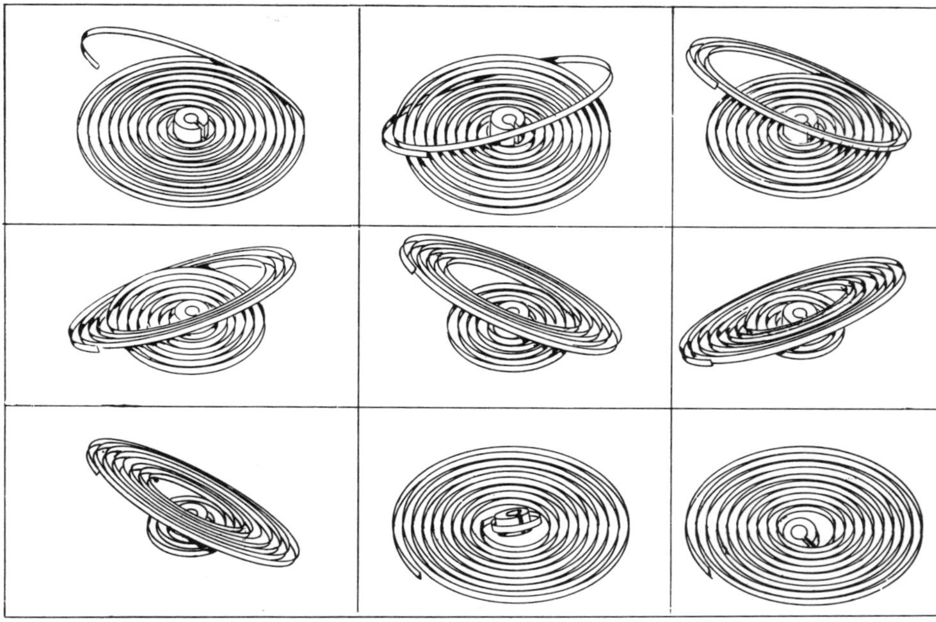

FIGURE 51: Just as there are exercises to develop skills in truing hairsprings bent out of round, there is a similar set of exercises to increase skills in correcting hairsprings out of flat. At first the spring is deliberately bent upward a half turn, then corrected. This chart shows only the intermediate steps. The complete exercise should show progressive bends of a half turn until the entire spring has been manipulated. Since the student or his master purposely imposed these bends on the spring, he knows how the bend was inflicted and where it is located.

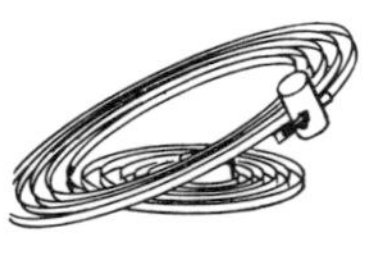

FIGURE 53

When viewed from its edge, an out-of-flat hairspring appears to be two springs, as in FIGURE 52. FIGURE 53 shows a simplified example, with the bend on the outermost coil. The bending point is at B, exactly 180 degrees from the highest spot between coils, A.

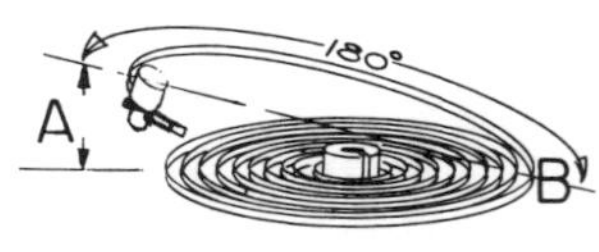

FIGURE 52

rate springs. Figure 53 shows the same type of bend, simplified, since it is on the outermost coil. The highest spot is at *A*. Spiraling inward from *A* toward the collet 180 degrees, the bending point is found at *B*.

There are three types of bends out of flat: those with coils twisted upward suddenly; those twisted gradually; and those with a "Breguet" bend, similar to the bend in a hockey stick, in which the ribbon of wire is bent against its width.

The best method of correction depends upon the cause of the bend. Usually out-of-flat bends in the last coil are caused when the spring gets caught in a part of the movement while the balance and bridge are being removed from the watch. In such a case, the spring's coils are *pulled* out of alignment. The logical correction, then, would be to *push* them back into alignment.

GRADUAL BEND

The gradual bend can best be corrected by holding the spring behind the start of the bend with sturdy but pointed tweezers and pushing down on the highest outer half turn with the index finger of the opposite hand, as in Figure 54.

If the hairspring is small, the finger method may not work too well. A pointed piece of pegwood or a needle set into a pegwood stick is often more successful to nudge the spring back into shape. This is shown in Figure 55.

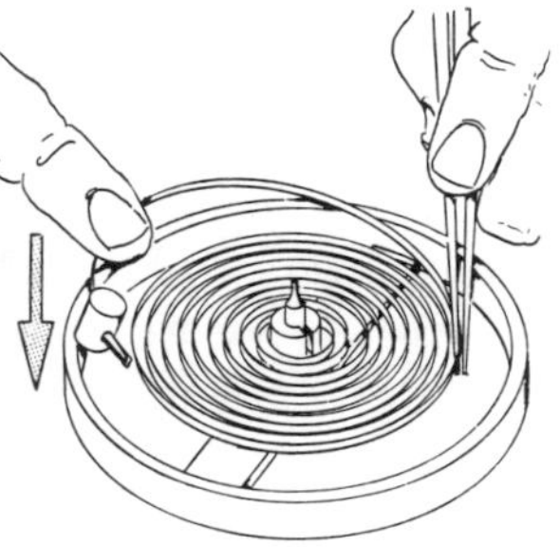

A gradual bend can be corrected by grasping the coil with tweezers just behind the bending point and pushing down with the index finger. A needle can substitute for the finger when working with a particularly small hairspring.

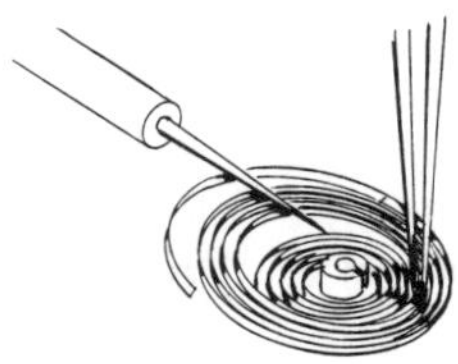

FIGURE 55

SUDDEN TWIST

The sudden twist (Figure 56) can also be corrected as in Figures 54 and 55, although some watchmakers would rather correct the kink with two pairs of tweezers. Figure 57 shows how one pair grasps the spring behind the kink closest to the spiral leading to the collet, while the other grasps the spring as close to the first tweezer as possible. Twisting the second tweezer in the direction shown by the arrow removes the kink and helps to level the spring. The vertical tweezer's points should touch the bench.

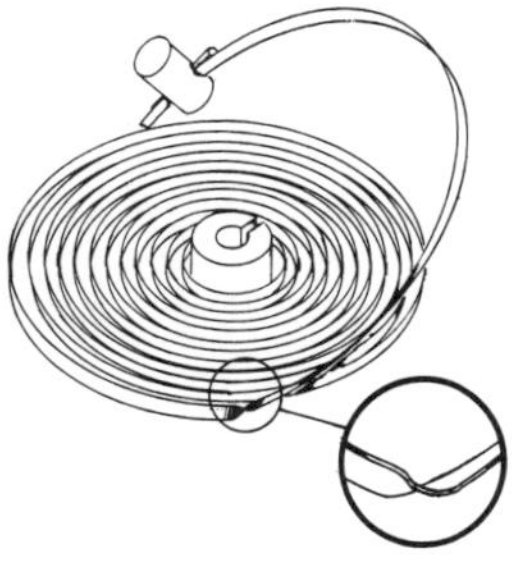

FIGURE 56

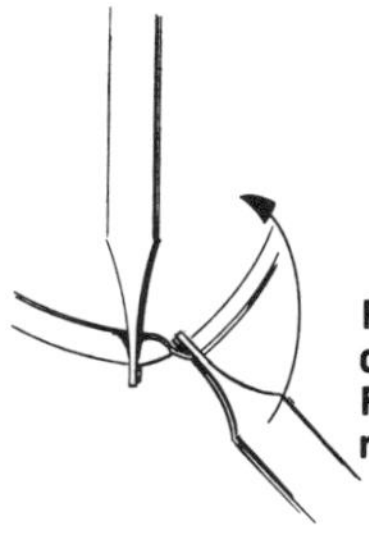

FIGURE 57

FIGURE 56 shows a sudden twist in an out-of-flat hairspring. Correct it as shown in FIGURE 57, by twisting in the opposite direction with two pairs of tweezers.

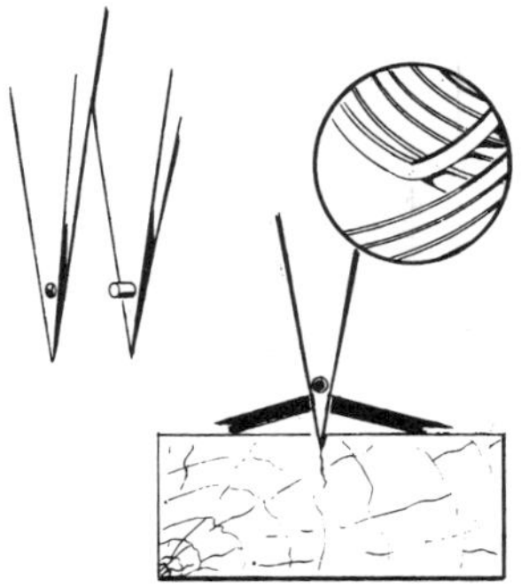

FIGURE 58

A Breguet bend should be handled with a special pair of tweezers fitted with a pin in one blade and a matching hole in the other. Grasp the hairspring at the top of the bend and rest it on a block of soft wood, across the grain. Gently press the tweezer points into the wood until the bend is straightened.

BREGUET BEND

Figure 58 shows a method of removing and correcting an inadvertant Breguet bend. Use a special pair of tweezers with a pin frictioned in one blade, fitting into a hole in the other blade. Grasp the hairspring at the top of the bend and rest it on a block of soft wood *across* the grain, tweezer points down. Now push down on the points into the wood. Usually this method of correction causes the spring to bend out of the round, but this is the easier of the two bends to correct.

To correct bends at the collet, use two pairs of tweezers. Grasp the coil close to the bend and push down (or up) on the opposite side. If the coil is bent at the spot where it emerges from the collet, place the collet over a tapered pin. Grasp the pin in a pinvise, and with another needle set into a pegwood stick, nudge the bend back to a level position. Remember, the closer you get to the collet, the finer the tweezer points required. You recall that bends and their correction are more critical at these central spots than on the outside coils.

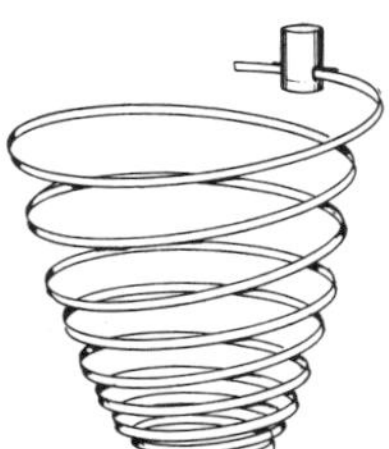

FIGURE 59

SPIRAL HELIX

Sometimes a hairspring is pulled grotesquely out of shape as in Figure 59. This spring looks hopeless. Actually it is one of the easiest to correct, but the method is always impressive to the student as he watches his master.

Such a spring has become a spiral helix, probably because the watchmaker carelessly pulled the balance bridge out of the movement, thinking that the balance, attached to the bridge by its hairspring, would easily follow. In these rare exceptions when it doesn't, the pivot has become lodged in the clogged jewel hole. The result is a spring such as you see in Figure 59.

Remove the spring from the balance. With one pair of tweezers, grasp the collet from the stud side down through the vortex of the spring in the following manner.

Insert one blade through the center of the collet and the other blade along the outside edge of the collet between the collet wall and the first turn of the hairspring.

FIGURE 60

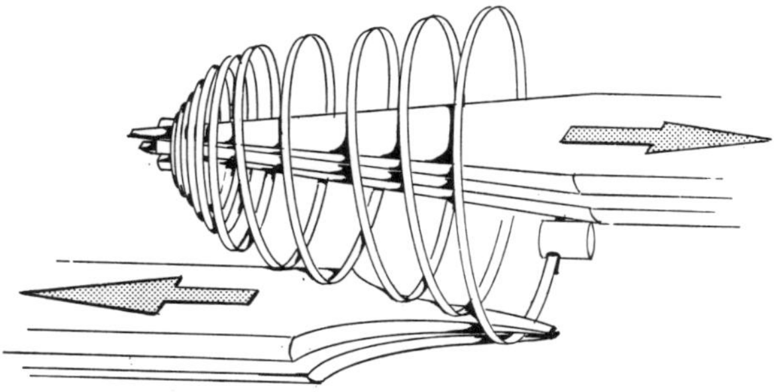

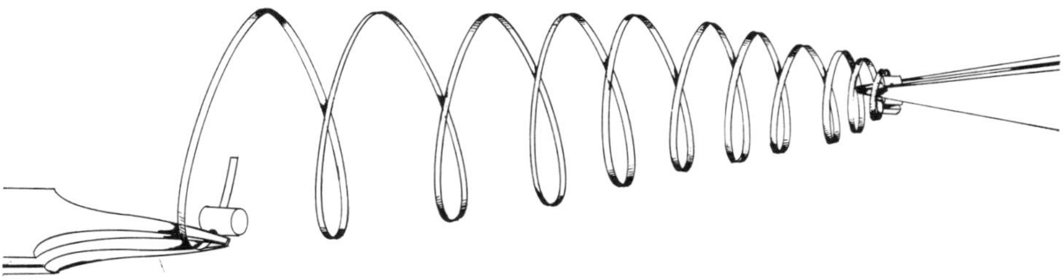

The spring in FIGURE 59 has been stretched into a spiral helix. Take it from opposite directions with two pairs of tweezers as shown in FIGURE 60, one down through the inside of the coil, grasping the collet; the other along the outside of the coil, grasping near the stud. Pull the tweezers apart along a parallel axis until they pass each other and stretch the spring in the opposite direction, as shown in FIGURE 61.

With a second, sturdier pair of tweezers, grasp the hairspring near the stud from the direction of the bottom of the stud, as shown in Figure 60.

Now pull the two tweezers apart along a parallel axis, as illustrated in Figure 61. The pull in the opposite direction must be greater than the distance the spring has been distorted, in order to exceed the point of elasticity and restore the spring to its original condition. Of course don't overdo it! Pull the spring apart in gradual, trial distances. Should you overcorrect the level of the spring, merely reverse the process until the spring is level.

WITHOUT REMOVING THE HAIRSPRING

Sometimes it is preferable to correct the hairspring while it is still in the balance, especially if the bends are situated in the first one-third turn outward from the stud. This requires a slightly different procedure.

When the bend is situated between the stud and the regulator, consider the stud as an anchor point, acting as tweezers would. Pinch the hairspring near the stud with a pair of tweezers tilted in such a direction to bring the spring up or down to correct its level.

This is a particularly good method of correction when the coil diametrically opposite the stud is interfering with the overlapping center wheel or rubbing on the balance arm, as shown in Figure 62.

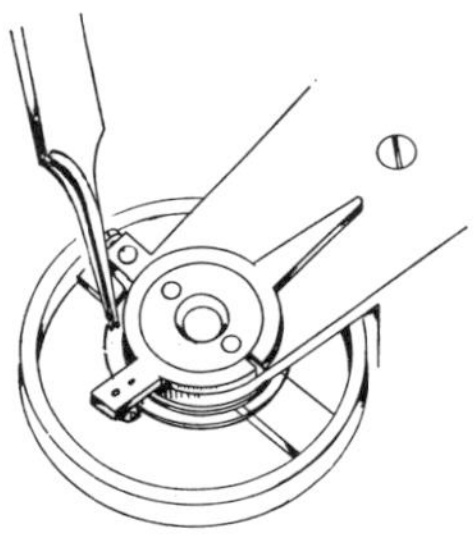

FIGURE 62

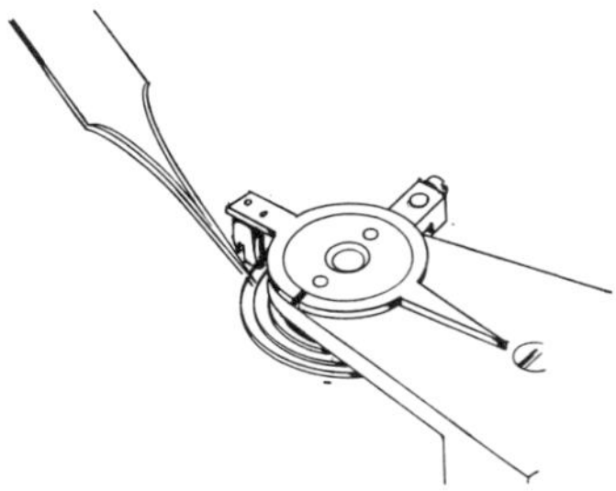

FIGURE 63

Sometimes it is advantageous to correct the hairspring while it is still in the balance. If the bend is situated between the stud and the regulator, as in FIGURE 62, consider the stud as an anchor point and pinch the bend gently with tilted tweezers to bring it back to shape. If the bend is outside the regulator as in FIGURE 63, consider the regulator the anchor point and pinch the bend in the same manner.

When the bend is situated just outside the regulator, consider the regulator as the anchor point. Pinch the spring with a tilted tweezer as shown in Figure 63.

The regulator can be shifted to act as a close tweezer to make corrections at opposite points. Merely move the regulator 180 degrees studwise from the bend. The closer the regulator is, the more effective is the correction.

Do not pinch the spring excessively between the stud and the regulator. This will only cause a secondary bend when and where the spring comes in contact with the center wheel or the bottom of the balance bridge, or when, in the opposite direction, the spring is given a secondary bend at the point where it bears on the toe of the regulator boot.

FIGURE 64
If a hairspring has two bends out-of-flat, treat each section of coil one at a time, working from the collet outward.

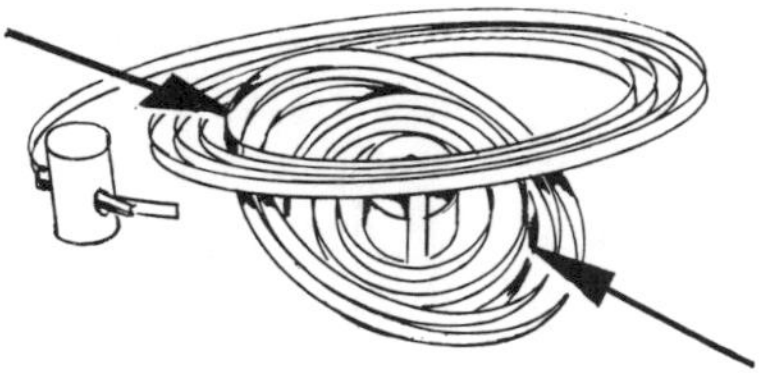

Always consider a hairspring as being in separate parts divided by the bends. If the spring has just one bend, treat it as being in two sections: one part is the main and correct part spiraling from the collet outward to the bend; the second is that with coils tilting at a different angle from the first part. Should there be two bends as in Figure 64, again treat each section separately, always using the section from the collet outward as a standard.

Proceed by correcting one section at a time. It may be more convenient to correct the outermost bend first, then work inward, but most experienced hairspring manipulators prefer working from the collet outward.

CHAPTER IV

VIBRATING THE HAIRSPRING

VIBRATING a hairspring is that operation in which the new spring's length is adjusted so that it causes the balance to swing a definite number of times a minute.

There are numerous methods of vibrating a new hairspring to a balance. To give detailed descriptions of each method would require a volume in itself. However, the more efficient and simpler methods will be described here.

Professional hairspring vibrators use a vibrating head with adjustable tweezers. This is illustrated in Fig. 65. The subject balance and hairspring are suspended from the tweezer until the lower balance pivot rests on the vibrator platform.

The platform upon which the pivot is made to rest is composed of a disc of transparent glass. Under the glass and set into the platform is a precision timed master balance and hairspring.

The suspended balance is adjusted so that its pivot is directly

Fig. 65. How sample balance is suspended over glass-enclosed master balance.

over the center of master balance and the arm of the balance under test is directly over and parallel to that of the master balance, as shown in Figure 65. Both balances are then shaken into motion by lateral movement of the platform and both balances are observed to see whether the tested balance gains, loses or is in unison with the master balance.

To the beginner, it is difficult to determine whether the balance is faster or slower than the master balance. With some practice, the eye will learn to make this distinction. An aid in this direction is to put both balances in weak motion. It is then more easily observed. If difficulty is still encountered in deciding whether the balance is slower or faster, observe the balance and you will notice that while both balances start off together and in step, this unison will soon be altered. One balance will continue to vibrate and the other will be seen to change until the arms are oscillating *against* one another. A little further on they will catch up to one another and again be in step. With a timepiece with a

Fig. 66. A balance scaffold and adjustable tweezer. Inset shows tweezer opened.

seconds hand, time this interval from the first vibrations when both are started in unison until they part and finally come together again. Make a note of the time it takes for the completion of this cycle. Shorten the hairspring and re-time this period. If the time period is shorter, then it is an indication that the shortening of the spring is aggravating the condition and it is a sign that the subject spring is too short.

On the other hand, if the interval becomes longer, it is an indication that you are proceeding in the correct direction and that the spring was slow (too long).

This process is continued cautiously until both balances keep in time for a considerable length of time or until both balances stop. Should the subject balance closely approach unison and the next changing of hairspring length produce a shorter period between the unison and out-of-step duration, it is apparent that the spring's length was over-compensated.

When both balances seem coordinated and remain so until they stop, the hairspring's length may be assumed correct. *The spot where the vibrating tweezer holds the spring becomes the point where it must lie between the regulator pins.* Extra hairspring length must be added to this point so that it may be studded.

For the watchmaker at the bench who chooses to vibrate hairsprings without elaborate equipment, there are a few good practical and simple methods which give accurate results. These methods depend on comparison to a reliable balance, "counting" the vibrations of the subject and utilizing the simple laws of harmonic motion with another pre-timed balance in a going watch.

The basic idea of comparing the subject balance with a pre-timed balance is very much the same as described previously and used by professional vibrators. However, the "master" balance may be just an old 18s watch stripped of all parts but the balance and bridge, or some similar simple device.

"Counting" is simply counting the vibrations of the subject balance for a given time.

The idea of using the laws of harmonic motion of the subject balance with a master balance will be described in detail.

(An old 16s or 18s movement, stripped down so that only the lower plate, balance and bridge remain, makes a good model with which to match the newly sprung balance. Of course, the watch should be timed and regulated to very close limits before it is dismantled in order for the balance to serve such a precise purpose.)

To use the comparison method, the subject balance's hairspring may be held in a regular bench tweezer or it may be suspended from a tweezer set into an adjustable platform or scaffold as shown in Figure 66. When the spring is held in the tweezer, the balance is brought directly over the master pre-timed balance so that the arms of both balances are directly over one another. Both balances are then twisted into motion at the same time and the subject balance is observed to see how long it will keep in unison with the lower balance.

If doubt exists whether the subject balance is slow or fast, time the interval between the start of both balances and the time that

Fig. 67. By enmeshing the screws of suspended balance with screws of the master balance and turning one with finger, they will start together.

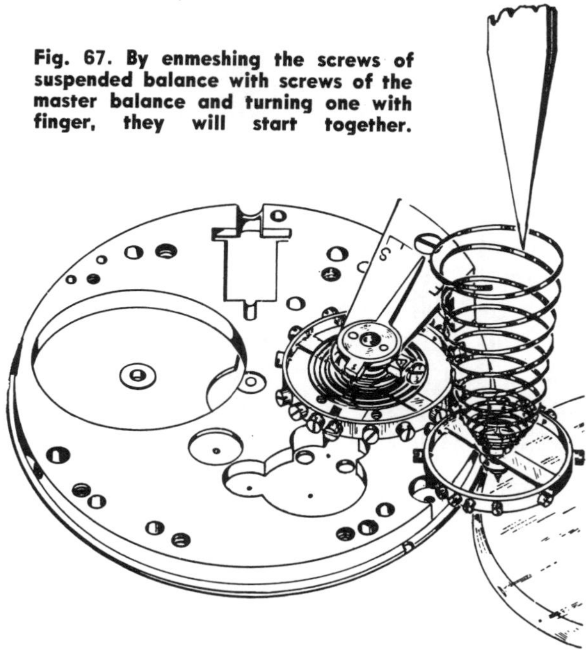

it gets out of and into step again. Then shorten the spring and time the cycle once again. As previously described in the paragraphs on the vibrating machine, if the second cycle is of a longer duration, it is an indication that you are proceeding in the correct direction and that the spring was too long. If the interval is quickened instead, the reverse is true and the spring must be grasped with the tweezers at a point nearer its end.

For some beginners in this division of watch repairing, the job of getting both balances to start together is sometimes a bit difficult. To overcome this, the balance under test is placed beside the master so that the screws of one balance enmesh with the screws of the other and fit similar to engaged cog teeth. Of course the arms of each balance are made to appear as a straight line as shown in Figure 67. In this manner if one balance is moved, it will motivate the other and both will be put into simultaneous motion. To do this, the index finger of one free hand slowly twists the master balance, which, having its screws enmeshed with the screws of the subject balance, will in turn rotate it as well. When the finger has moved the master balance about a ¾ turn, the finger is suddenly removed, releasing the master balance. At the very same moment that this is done, the other hand moves the subject balance sufficiently aside to have its screws clear those of the master balance. Both balances will therefore start off alike.

As they swing back and forth, the arms of both wheels will seem to part and come together again. Thus the length of time this action remains in unison is observed and the spring's length adjusted accordingly. Of course when the subject balance is moved slightly aside, a smooth surface such as a flat watch glass is provided for its lower pivot. The balance should be suspended so that its lower pivot barely rests on the surface chosen.

Another method of bringing the spring's length to size and time is to count the number of vibrations of the balance in a given time, usually for a minute. This method is not as accurate as matching the vibrations of the balance with another, but many watchmakers are able to master this system with remarkable results. A watch with a large seconds hand or one with a sweep seconds hand is used as the timepiece with which the counted vibrations are directly checked. In fact the counting is done directly upon the crystal over the watch dial to facilitate the observation and counting.

Fig. 68. A watch with a sweep second hand can be used to count vibrations of trial balance as alternate method.

The balance with is new spring is grasped with the bench tweezers and suspended over the center of the watch crystal by bracing the fingers of the hand holding the tweezer in such a manner that its lower pivot barely touches the crystal as shown in Figure 68. With the other hand the balance is twisted around so that when it is released it will have an active motion. When the balance is released and it starts its swings it will be drawn up off the crystal and then down again with an audible tap by the descending lower pivot. These taps are then counted beginning when the second hand on the watch under the crystal *passes* (not reaches) an easily remembered part of the dial, such as the 60-second mark or any of the quarter marks.

Since these taps represent only alternate beats, they are infrequent enough to be counted. The total number of taps in one minute is then multiplied by two. In most watches, there are 300 vibrations a minute. If the spring is of perfect length then there should be 150 taps per minute. If the total number of taps on the

Fig. 69. Another method is to suspend the balance and to observe the vibrations, timing them by the sound of a going watch held against the ear.

crystal by the lower pivot in one minute is more than 150, it is an indication that the spring is grasped at a point too far from its outer terminal. If there are less than 150 in a minute, then the tweezer must grasp the spring at a point farther from the outer end of the hairspring.

Sometimes the motion of the balance falls off to such an extent before the minute is up that the taps upon the crystal become too weak to be heard. In that case the balance also has a very short arc and it then becomes possible to pick up the count of the alternate vibrations by visual observation. That is, the arcs are short, slower and easier for the eye to follow and thus pick up a count. Or the taps heard may be timed for a half minute period. As with the acquisition of many other watchmaking skills, practice is essential.

Another method is the system in which the balance is suspended and the vibrations observed as explained above. However, instead of the pivot resting upon the watch, the going timepiece is held against the ear. This is shown in Fig. 69. If the suspended balance oscillates in harmony with the watch held against the ear, the ticks will seem to come from the suspended oscillating balance. This, of course, is an indication that the spring's length is correctly adjusted.

If the spring on the suspended balance is longer or shorter than it should be, the speed of this balance will, of course, be out of harmony with the watch against the ear. This will manifest itself in the fact that the ear and eye will fail to associate the sounds of the ticking balance with that of the out-of-time oscillations of the suspended balance. As soon as they re-enter a phase of unison, the association of sight and sound will again take place. The longer the tested balance and the audible watch remain in unison, the more accurate is the vibration of the tested balance spring. A

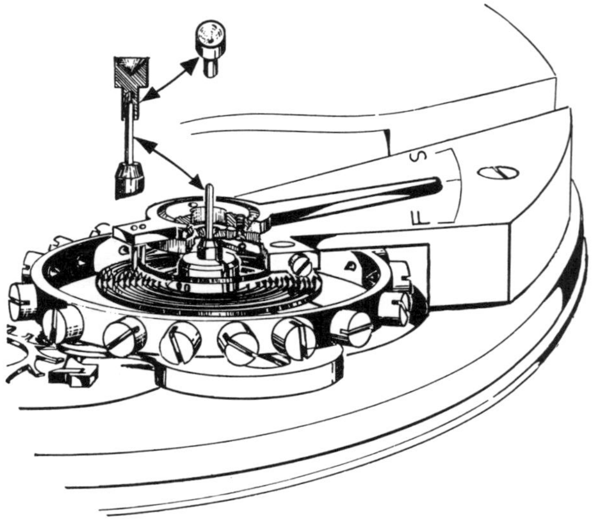

Fig. 70. Device for testing balance as described in a Swiss journal, employs specially-made device pictured here and used on precision-timed watch which is adapted for its use.

few trials with an assortment of balances will reward the operator with the desired efficiency.

HARMONIC MOTION

It is known that if a pendulum was to stand close to a going clock whose pendulum's length approximated its clock-less neighbor, the idle pendulum would soon be set into motion. If the clock-less pendulum is shorter or longer than the clock pendulum it will stop a short time after it has started and then start all over again, stopping soon after that. These cycles will recur with some degree of regularity. However, if the pendulums are almost exactly

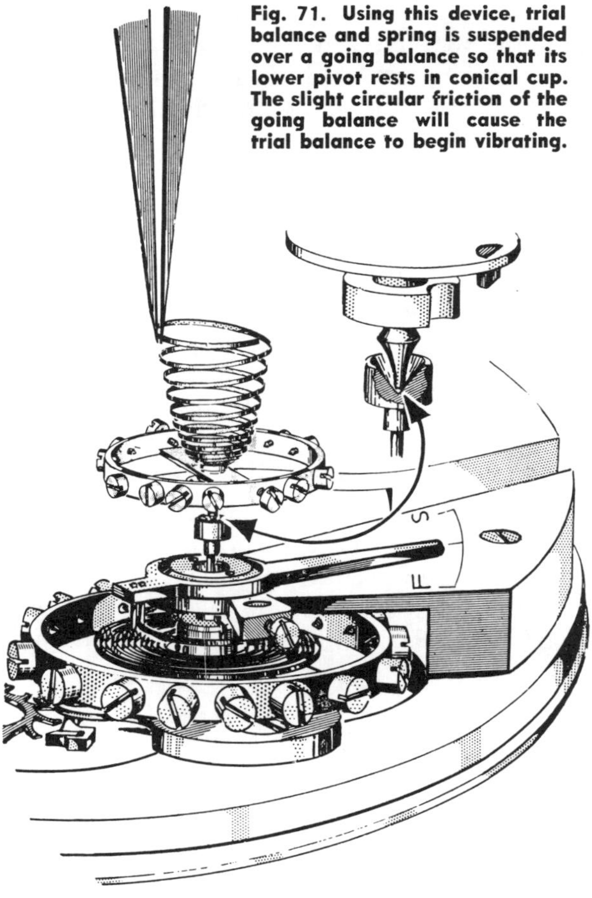

Fig. 71. Using this device, trial balance and spring is suspended over a going balance so that its lower pivot rests in conical cup. The slight circular friction of the going balance will cause the trial balance to begin vibrating.

alike in length, the free pendulum will remain in motion a longer period of time. The closer the pendulums match each other in length, the longer the free pendulum will oscillate. Breguet made such a clock utilizing this principle.

Utilizing this same principle, it is reasonable to expect that two balances, brought close enough to one another would react in the same way. By the manner in which the escapement-less balance reacts, its timekeeping values may be judged accordingly. To utilize this idea, the Swiss journal describes a method in which it is possible to bring the balance with the newly-fitted spring directly over the balance of a going precision-timed watch. The lower pivot of the balance under test is rested in a countersink above the elongated upper pivot of the heavy balance of the going watch below.

To produce such a device, a good grade large-sized watch with a heavy balance is obtained. It should be in going condition. Then the balance staff is re-pivoted or a new staff is made so that its upper pivot is about as long and almost as thick as the fourth wheel pivot. The cap jewel is removed from its setting and another hole jewel is substituted for one to fit the new pivot. A brass socket, similar to that on a seconds hand is made to fit over this protruding pivot. However, the top of this socket is cut so that it has a conical countersink, similar to the balance staff nut on a cheap alarm clock. This unit is shown in Fig. 70. After it is fitted and the watch is put in good running condition, it is re-timed and regulated to compensate for the additional weight upon the balance.

To use this device, the balance and spring under trial is suspended over the going balance so that its lower pivot rests in the conical groove in the socket atop the going balance. This is shown in Fig. 71. The suspended balance will soon start to vibrate, impelled by the slight circular friction caused by contact of the socket with the lower pivot. If the suspended balance has an erratic or irregular motion, it is an indication that the hairspring is either too long or too short. The tweezer may then grasp the spring nearer or further from its outer end. As in the case of the pendulums, if the balance starts off and soon stops, it is an indication that the timing of these balances is poor. If the suspended balance picks up motion and remains that way for a long period of time, it is obvious that the length of the hairspring is fairly correct. If the hairspring is of correct length, the motion of the balance picks up considerably and both balances will oscillate in harmonious motion almost indefinitely.

When using this method the balance may be suspended so that

the pivot always remains in contact with the socket and the spring may be supported by the tweezers held in the hand. Of course, it would be better if the spring were held in the small mechanical tweezers set into a scaffold such as shown in Fig. 66.

CHAPTER V

STUDDING THE HAIRSPRING

AFTER the hairspring has been vibrated and its working length ascertained, it must be fastened to its stud at a point near its outer terminal. This operation is called "studding." When the vibrating point has been found, it will eventually become the part of the hairspring which is gripped between the regulator pins. Therefore an extra length or section must be added beyond 'the vibrating point so that the end of this length may accommodate the stud as shown in A, Fig. 72.

If the hairspring is to remain a "flat" one, about ⅜ of a coil is added to the vibrating point, depending upon the style of the balance

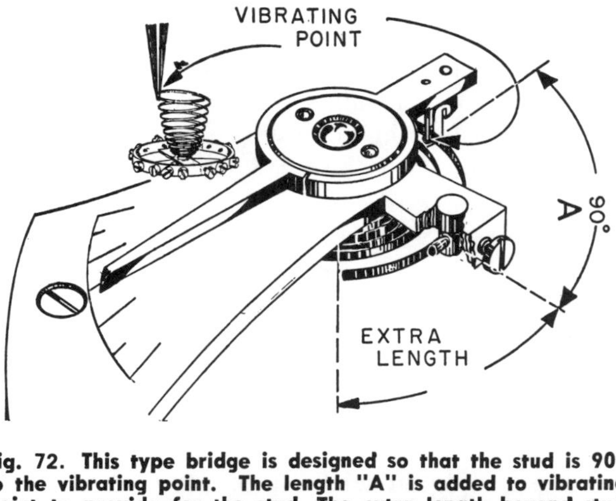

Fig. 72. This type bridge is designed so that the stud is 90° to the vibrating point. The length "A" is added to vibrating point to provide for the stud. The extra length beyond stud is left for possible future adjustment.

bridge. In Fig. 73 is shown a bridge where the stud hole is 45° beyond the (vibrating point) regulator. Such a bridge does not require more than ¼ extra coil beyond the vibrating point. The bridge in Fig. 72 has the studhole at right angles (90°) to the regulator. Such a bridge would suggest that the spring extend

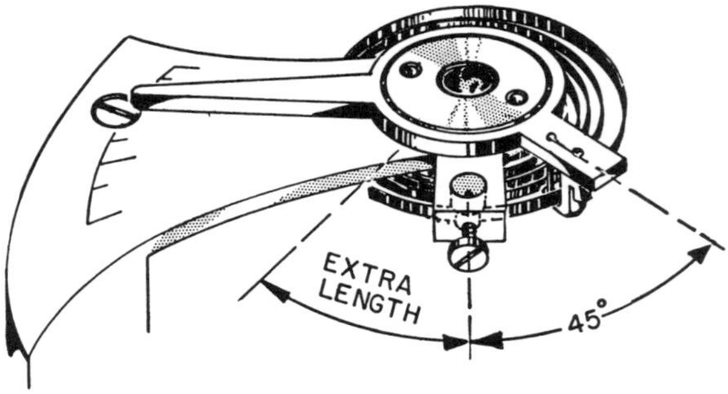

Fig. 73. This type bridge is designed so that the stud is only 45° from the vibrating point.

about ⅜ beyond the vibrating point. This would leave ⅛ of a coil protruding beyond the stud. The purpose of this surplus is that it acts as a reserve length to be utilized in case the vibrating was miscalculated or for the later re-adjustment, should it be needed.

In a Breguet hairspring, no extra length is advisable beyond the stud because any bit of spring extending freely out of the stud will only catch upon the section of the spring which ascends from the lower to the upper coil. This fault is shown in Fig. 74.

The actual operations of removing the old section of spring from the stud is very much the same as that employed in clearing the collet hole. However, the stud is easier to manipulate because it is at the outermost part of the hairspring and therefore more accessible. Holding the stud for the removal of the old pin and section of the spring may be done with the stud held in the bridge to which it will eventually be secured (Fig. 75), held in a pinvise (Fig. 76), or in a special studding "table" shown in Fig. 77.

The balance bridge may be used to hold the stud for the studding job when it is large enough to be handled or when the stud has a very irregular shape as shown in Figs. 75, 83, 84. Otherwise the stud may be held in the devices previously described.

EJECTING THE OLD PIN

The old stud pin is ejected with the point of a steel needle pushing the pin out from the thin end toward the thick end. When both ends of the pin are flush with the sides of the stud hole, it may be

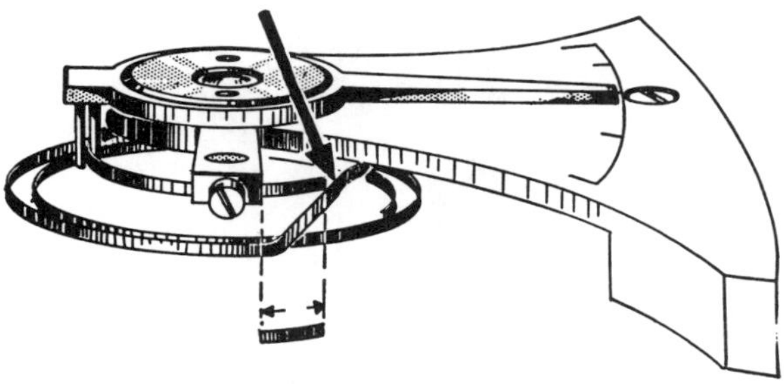

Fig. 74. In a Breguet hairspring, no extra length outside the stud is advisable because this may touch the rise in the overcoil as shown by the arrow.

assumed that the thin end is at the side of the stud from which the spring end emerges. When small bits of the hairspring stud-pin still extend from both sides of the stud, it may be ejected by bracing one point of a strong pair of tweezers upon the thin end of the pin and the other tweezer point braced against the side of the stud. Squeezing the tweezer points together will loosen the stud pin so that it can be removed as shown in Fig. 78.

The studding table shown in Fig. 77, while used primarily in mass production work and by professional hairspring vibrators is a very handy tool, making the studding job much easier. The table

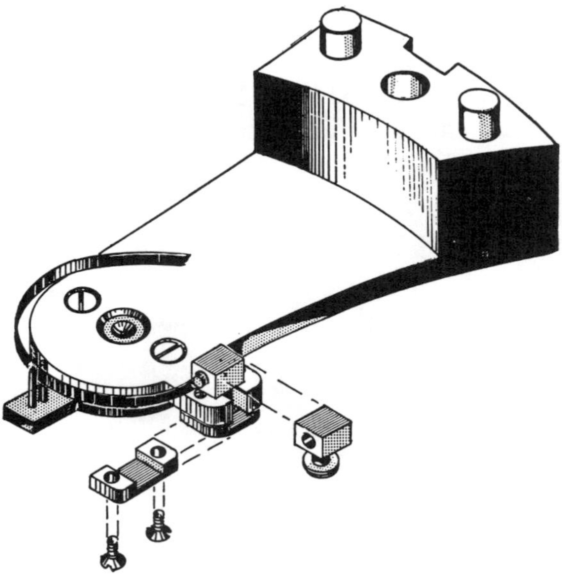

Fig. 75. The studding may be done in the bridge, especially if the stud is an odd shape. The stud shown here is mushroom type.

consists of a platform "A," a steel piece "B" dovetailed into the front edge of the platform into which are cut a few notches of varying sizes to accommodate studs of different dimensions. "C" is a strong steel spring which is used to clamp the stud against the notches in the piece "B." The spring is fastened to the table by two screws or one steady pin and a screw. The lever "D" pushes the spring outward and open to release the stud or to provide space when securing it. In this way, the stud, head down, is ready for the operation of having the spring threaded into it and fastened.

HAIRSPRING STUDS

There are various types of hairspring studs, coming in many shapes and sizes, as well as a variation in the means by which they

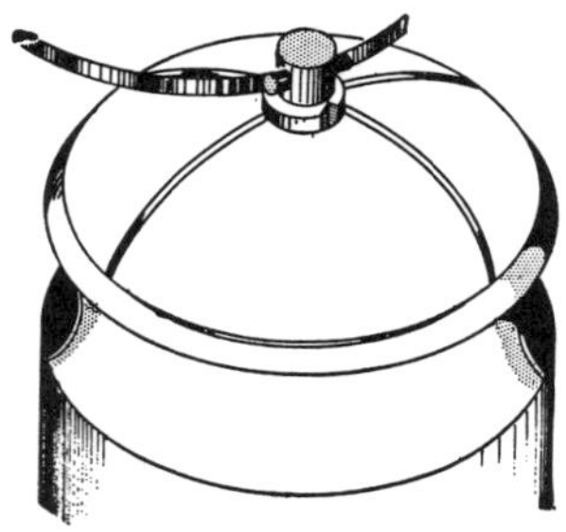

Fig. 76. Stud held in a pinvise.

become secured to the balance bridge. Some of these are pictured in Figs. 79, 80, 81, 82, 83, 84, 85. In Fig. 56 is shown a conventional round type of stud. This is by far the most popular style used. The longitudinal "V" groove is used to seat the stud (set)

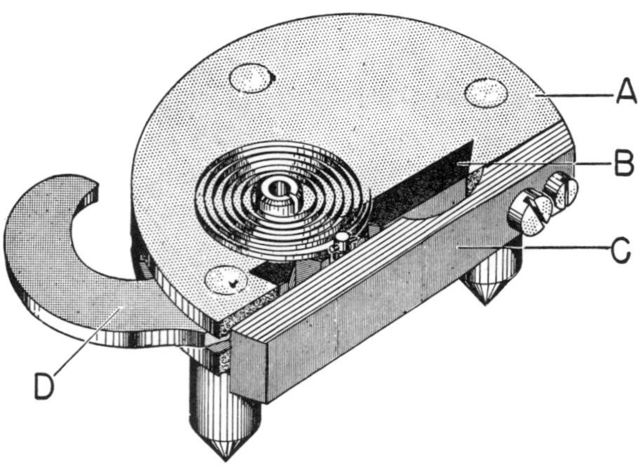

Fig. 54. Studding table is used here to hold the stud while it is pinned.

screw keeping it from shifting around in its hole in the bridge and causing the hairspring to become de-centered.

A form of triangular stud is shown in Fig. 80. This is used in some American and Swiss watches. The triangular shape seated in a form-fitting hole in the bridge prevents it from shifting.

Figure 81 shows a pentagonal stud. Its shape is designed to keep

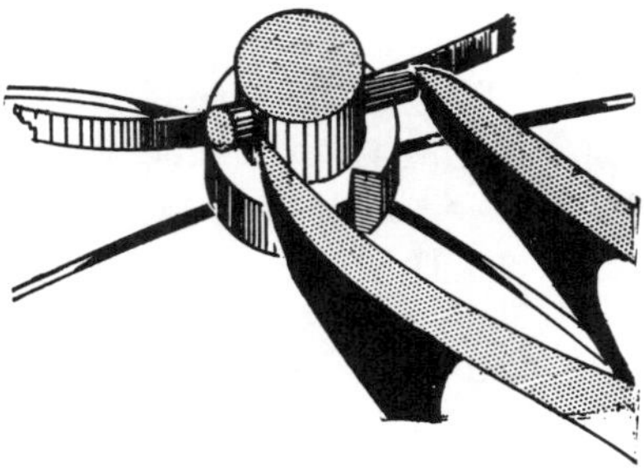

Fig. 78. Removing the old stud pin.

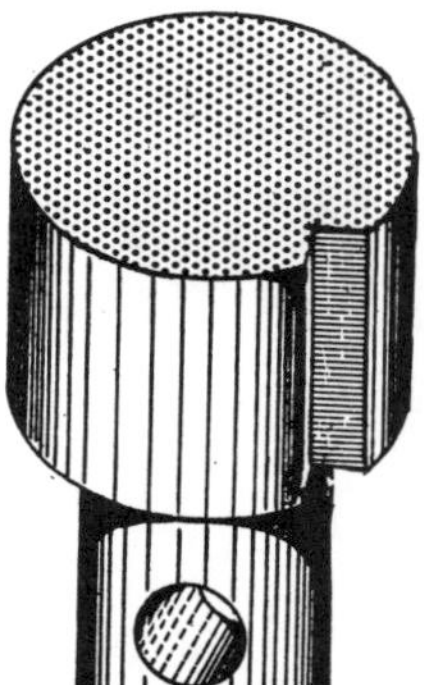

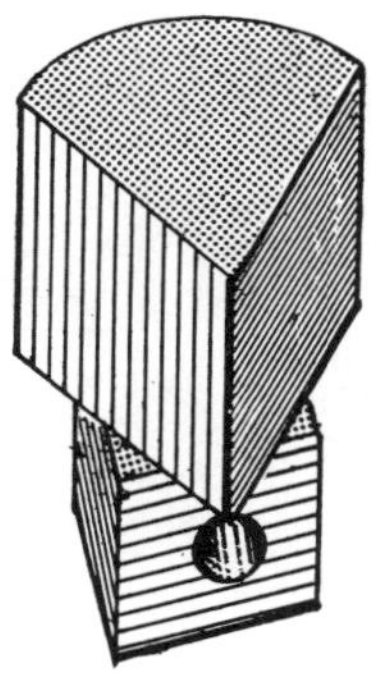

Figs. 79 and 80, above, and Figs. 81 and 82 below show the conventional type studs.

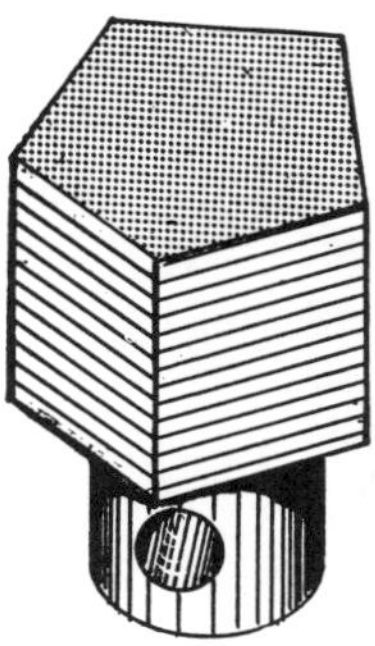

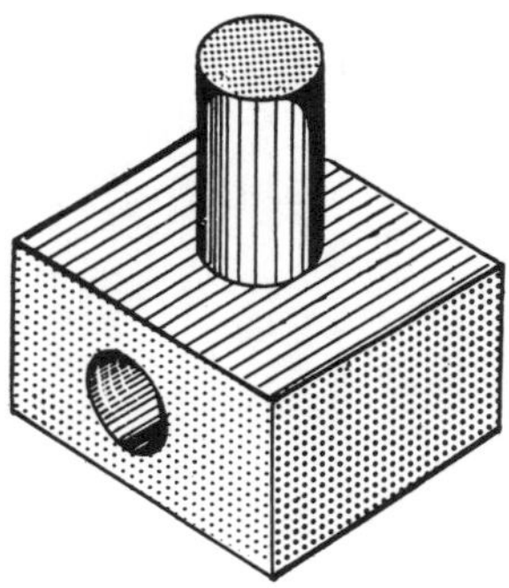

it from shifting and its angle, blunter than that in Fig. 80 assists the stud screw in seating it better in the stud hole.

The stud in Fig. 82 is one met with in cylinder and similar watches. It is called a "Friction" stud because it is secured to the bridge by its thin post which fits friction tight in the hole in the bridge. It has no stud set-screw to bind it.

Studs in Figs. 75, 83, 84, 85 fall into the category of *floating* studs—so called because they are attached to the balance bridge (with the hairspring) after the balance is in the movement and the bridge secured in place. The stud, attached to the spring is eased into its "dock" or "pier" and then locked in place with a set screw or a clamp screw. Figure 83 shows this floating stud attached to the bridge. A screw passes through the bevelled hole in the top of this stud and screws into the threaded hole in the bridge. This type is called a *Right Angle* and sometimes a *Box* stud.

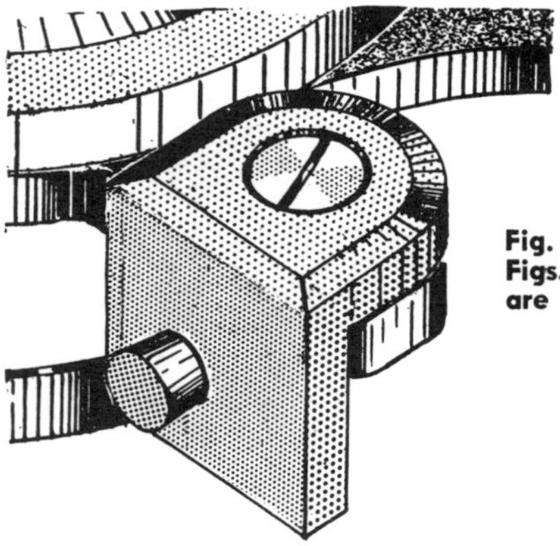

Fig. 83, at left, and Figs. 84, 85, below, are floating studs.

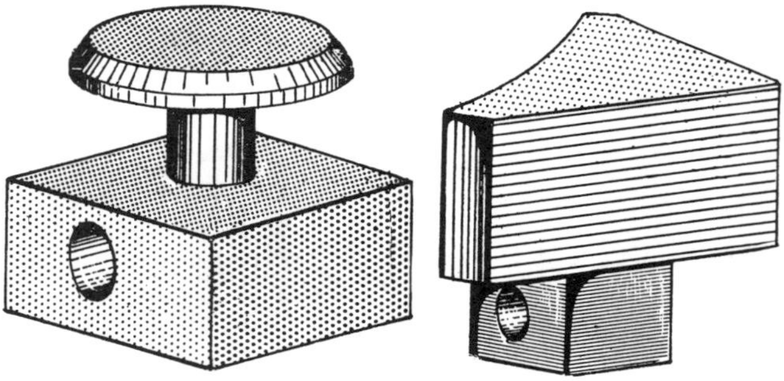

The "Mushroom" or "Collar-button" stud shown in Fig. 84 is shown in detail in Fig. 75. This stud fits into the bridge by having its neck go into the slot in the bridge. It is *clamped* in place by a stud cap and the two screws fitting over the "mushroom" of the stud. Figure 85 shows the Waltham type of stud. A clamp screw secures it in place and a part of this screw-head acts as the third or retaining wall (of a triangle) to hold the stud in place. (See Fig. 11, Book II, Chapter 2.)

The round or cylindrical stud is the most commonly met with and will be used to demonstrate this operation. After the stud hole is cleared, the spring is threaded into the stud, bearing in mind that both the spring and the stud are upside-down and that the direction of the spring must be checked.

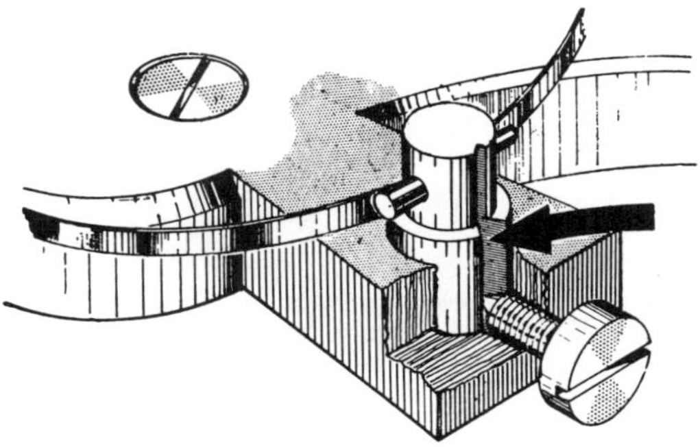

Fig. 85. The hairspring should be studded so that the groove (arrow) is facing the screw. This enables it to be fastened without danger of shifting.

Caution must be observed that the hairspring is threaded into the stud so that when it is placed back into the bridge, the groove in the stud faces the stud screw as shown in Fig. 85. In this way the stud screw will enter the vertical groove and prevent the stud from shifting in its hole. The stud is better secured because of the greater surface contacted by the tip of the screw.

If the stud is placed on the hairspring incorrectly as shown in Fig. 64, the stud screw has little grip on the smooth, round surface of the stud and may shift. This will cause the hairspring to be un-centered which in turn will cause the coils of the spring to crowd one another and the timing will be off.

After these precautions have been observed, the exact distance from the vibrating point to the outside of the stud (on the far side) is measured. A kink is placed in the spring at the point it emerges from the stud as shown in Fig. 89. This is done so that the exact studding point will not be shifted and lost during the studding operation, thus detracting from the exactness of the vibrating operation.

Taper pins similar to those used in the colleting operation are employed to wedge tight the stud and hairspring. This is inserted into the stud from the direction of the regulator pins toward the end of the hairspring. Of course the pin is placed inside the stud hole beside the hairspring so that it is *outside* the hairspring coils

Fig. 87. This hairspring was studded incorrectly. The groove (upper arrow) is opposite the screw and point (lower arrow) does not grip stud securely.

and facing the stud screw. The stud pin is then made secure by bracing a strong tweezer point on the thick end of the stud pin and the other tweezer point against the opposite edge of the stud, reversing the process shown in Fig. 78. This process is shown in Fig. 88.

Fig. 88. Inserting taper pin to fasten spring, one point of the tweezer is placed over the thick end of the pin and with the other point braced against the stud, the pin is squeezed in tight.

Since the stud is at the end of the spring and more accessible to manipulations, the precautions taken with the pin and spring during the colleting operation do not have to be repeated here, such as preparing the pin and using a sample length of hairspring for trial fittings.

In studding, the pin ends sticking out of the stud may be broken off by bending them outward. However, precaution should be observed that the kink made to mark the exact studding point has not shifted.

CHAPTER VI

CENTERING THE HAIRSPRING

FOLLOWING the studding operation, the spring must be centered so that when the spring is attached to the bridge, the collet center will lie directly over the balance hole jewel. To center a flat

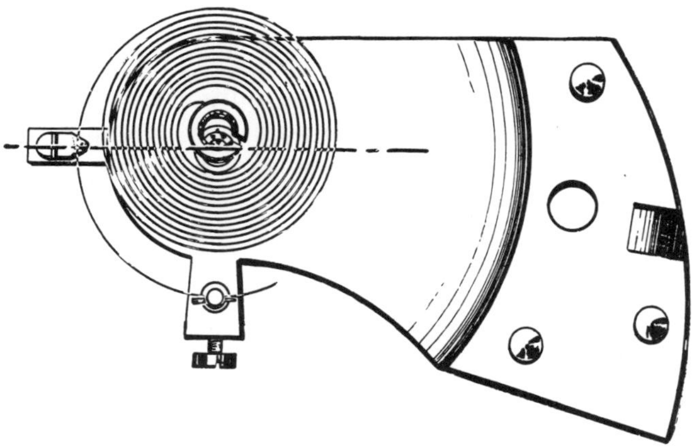

Fig. 89. If the spring is off-center as above, correction is affected by a number of bends in order to contract the curves.

hairspring, it is placed in the balance bridge *without the balance,* its stud screw fastened and the spring placed in the regulator. At first, the spring may be off-center as shown in Fig. 89. In such a case, the correction may be affected by a number of bends either to contract the curve of the last coil or a series of similar manipulations, usually to the last coil.

The truing in the round should not be considered completed until the spring and collet center are concentric with the balance jewel and the hairspring rests midway between the regulator key and the regulator pin as shown in Fig. 90.

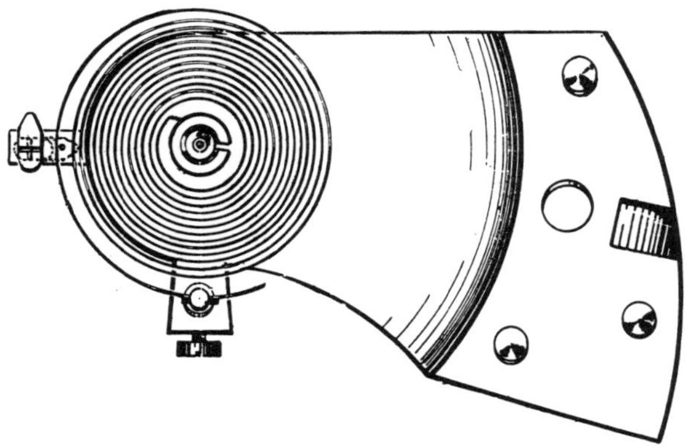

Fig. 90. Trued hairspring has spring and collet center concentric with balance jewel and the hairspring resting midway between regulator key and pin.

Sometimes a bridge is encountered where the distance from the center of the balance hole jewel to the stud ("A," Fig. 91) is greater than the distance from the balance hole jewel to the regulator pins ("B," Fig. 91). Since the outer ends of the regulator move in circular arcs, the section of the last coil of the hairspring that lies in the regulator must also conform to this shape. This is necessary so that when the regulator is shifted, it will not push the hairspring off-center. Therefore, when such a bridge is met with,

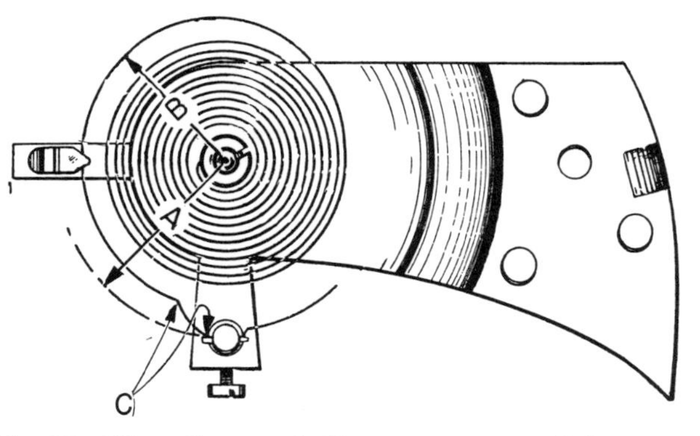

Fig. 91. When distance A (from center to stud) is greater than the distance B (from center to regulator pins) bends at C must be made so that last position of coil remains concentric with regulator arc B.

it will be required to make an adjustment close to the stud so that it will look like that shown in Fig. 91. This is done by bending the spring close to the stud at the points indicated by the arrows at "C."

It is taken for granted that all considerations as to the size of the spring in relation to the available room in the balance bridge have been observed and that the inner coils do not crowd the side of the regulator pin closest to the collet. (See Fig. 7, section, "Size of the Hairspring.") When the spring seems centered on the bridge, it is then removed and staked to the balance, making certain that the hairspring is replaced at the proper beat-point. The hairspring and balance are then attached to the bridge and minor and final adjustments to the hairspring may then be made.

CHAPTER VII

FORMING A BREGUET HAIRSPRING

A BREGUET or overcoil hairspring is one where the last coil is raised above the level of the other coils of the spiralled spring and made to curve inward.

The purpose of such a spring is to enable the balance to perform each oscillation in equal time throughout the run of the mainspring. When a hairspring achieves this effect it is called *isochronal*. To understand this we must first look at the flat hairspring. This

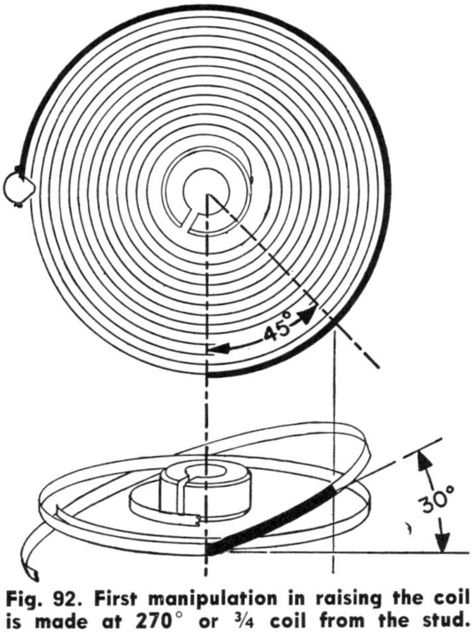

Fig. 92. First manipulation in raising the coil is made at 270° or ¾ coil from the stud.

"breathes" along a line from a point midway between the regulator and the stud, through the balance jewel center. The coils alternately crowd together towards the regulator and stud, and then expand diametrically opposite. This *torque* causes the balance staff to be unduly thrust against the side of the hole jewel, thus creating additional friction and robbing the balance of freedom of motion. The watch will thus gain time when it is fully wound and lose time when the mainspring is almost run down. When a watch behaves in this manner we say that it lacks *isochronism*. That is, the balance unit does not keep the same time when the mainspring is fully wound as when it is almost unwound and the balance arcs are short.

The Breguet hairspring, when its overcoil is properly formed, permits the spring to dilate and contract evenly all around, similar to the ripples created by a pebble dropped into the water. When the hairspring acts this way, coupled with a poised balance, it will eliminate the fault-causing side-thrust (torque) on the balance staff pivots. This in turn will diminish the errors in time-keeping inherent in the flat hairspring.

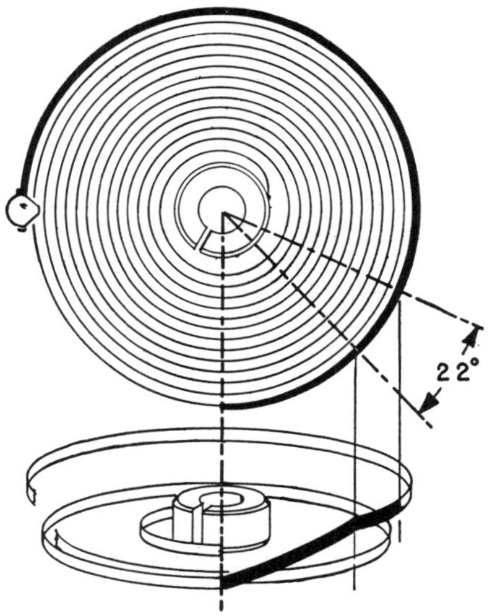

Fig. 93. Incline consuming only 45° of last coil, the spring is then levelled off and continues along the same spiralled path for another 22°.

All watches are designed to carry either a Breguet or a flat hairspring (although it is possible to form a Breguet and to fit it to a watch designed for a flat hairspring).

Should the sample be missing, it is a simple matter to determine what type of hairspring the watch requires. This is done by observing the balance bridge and regulator and referring to Figures 2, 3 and 4. It will be noticed in the flat hairspring that since all the coils lie in one plane, the regulator must reach the outermost coil and thus is longer. Generally, the key holding this spring is shaped like a boot as shown in Figure 2.

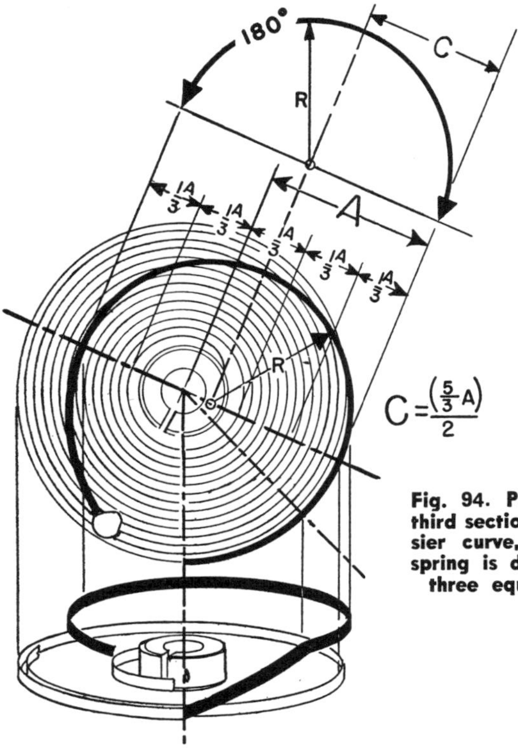

Fig. 94. Planning the third section of a Lossier curve, radius of spring is divided into three equal parts.

A watch requiring a Breguet spring has the regulator pins closer to the balance bridge and its stud hole is correspondingly closer.

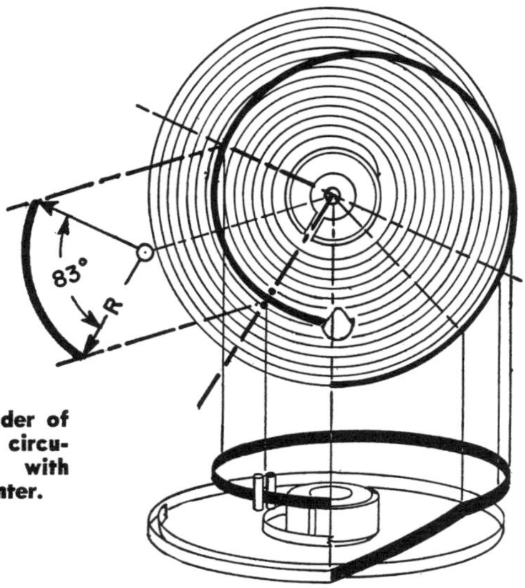

Fig. 95. Remainder of spring must be circular, concentric with the collet center.

This is so because the last coil is brought above and then coiled in a radius smaller than that of the overall size of the spring. Generally the regulator has no key but instead, two straight thin pins. one on each side of the overcoiled spring.

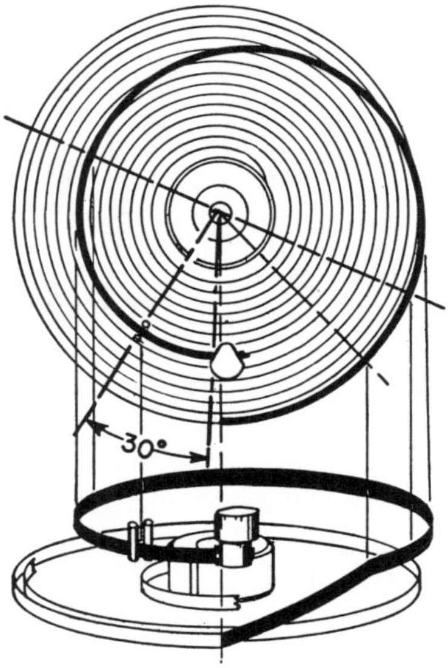

Fig. 96. From regulator pins 30° is added, using the collet center again as pivotal point.

When a hairspring is to become a Breguet, it is prepared in the same way as a flat spring. It must be colleted, timed and fastened to the stud, using the methods and precautions previously suggested. After the studding operation has been performed, its last coil is then raised and shaped to conform to certain theoretical and physical requirements.

There are a few accepted methods of raising the last coil. One is achieved by twisting the spring so that the last coil appears tilted and then leveling off the outer end by the same process. Another is to pull the spring upwards, while a third method bends the spring in a goose-neck fashion with a bend similar to that in a hockey stick.

Before actually making the bend, some thought should be given to where the first bend is to be made, the exact height of the overcoil and its shape.

The spot where the first bend is to be made depends a great deal upon the shape of the overcoil and its length. This may be governed by the physical restrictions of the balance spring and its proximity to the center wheel, the distance from the regulator pins to the balance hole jewel center (effective radius of the regulator) and whatever desire the operator may have to adhere to the requirements of the theoretical isochronism.

Great stress is laid upon these isochronal curves by books on springing and adjusting, when in practice it is difficult to achieve the isochronial adjustment by shaping the last coil only. This is due to faulty escapement construction and adjustment, too wide an arc of regulator movement, and the physical restrictions of the movement itself. Therefore, the instructions to follow will deal only with the operations of raising and forming the overcoil. However, attention will be drawn to one theoretical curve as a general model to follow.

In some books are shown about 200 assorted shapes of overcoils, each serving as a model, and each conforming to the requirements of an isochronal spring. Since so many different curves and shapes are illustrated, almost any curve made of the last coil will closely resemble one of these. If the hairspring is shaped so that it will vibrate evenly all around and its coils are concentric, it will conform to one of the standards and, within certain limits, serve its purpose fairly well.

In Figures 92 to 96 is shown a most popular type of Breguet curve. This was designed by Lossier after the theories of Prof. M. Phillips, published in 1861. For the purpose of implanting this in the reader's mind, it would be well to copy these drawings.

The first manipulation is made at 270° (¾ coil) from the stud. Because this will become smaller in diameter, the stud will eventually be close to the first bend. A study of these figures will reveal that the incline is supposed to consume only 45° of the last coil (see Fig. 92). Then, as shown in Fig. 70, the spring is levelled off and continues along the same spiralled path for another 22° before the curve is bent inward. Thus 67° of the last coil follows the same spiral path as the body of the spring of which 45° is the climb and 22° the "level-off." From this point onward, the last coil takes on a new path. Instead of following the spiral path it becomes circular; that is, for the next half coil it becomes a semicircle and at its terminal, amounting to about 103°, part of another circle. This is shown in the succeeding illustrations.

Figure 94 shows how the third section is planned. Here, the radius of the spring, measured from the end of the levelling-off point to the center of the spring, is divided into three equal parts (A, Fig. 94). Two of these thirds are added along a diametrical line across the spring. Thus five-thirds of this radius (A) equal the diameter of the third section. Therefore, to find the center of the third part of the Lossier curve, 5/3-A is divided in half. This will give the radius C. A semicircular arc (180°) is drawn as shown in Figure 94. This, added to the first and second section of the spring now has the appearance as shown in this drawing.

From this point, the remainder of the spring must be circular, concentric with the collet center. With the collet center as a pivotal point, 83° of arc is added to the overcoil. This brings the spring up to the point it enters the regulator pins as shown in Fig. 95. From here to the stud, 30° is added, using the collet center again as the pivotal point (Fig. 96). Thus the total of all the sections of the overcoil can be summed up as follows:

45° for the riser
22° for levelling off
180° first bend inward
83° next section to regulator pins
30° to the stud

360° total or one full coil

Of course, since the arcs do not all have the same center we cannot consider the coil as having a continuous or circular curve. Nevertheless, the spring should end with the stud on a radial line which also crosses the spring where the first (upward) bend was made.

It must be remembered that what has been described is the Lossier coil and its simpler technical details. In practice, however, it is not always possible to carry out these requirements. To do so would require perfect matching of the hairspring diameter in relation to the regulator radius, regulator arc and stud placement, and height of the Breguet. However, it is advisable to use it as a good model along which to pattern the new overcoil.

Before bringing up the spring to a new level, some consideration must be given to where on the new spring the first bend upwards should be made. Another factor to be observed is the height of the new level.

The first bend upward is made about ¾ of a coil from the stud. Because this last coil will become smaller in diameter than the main body of the spring, this section will become almost a full coil after it is raised and curved to a smaller diameter.

The height of the overcoil should be ascertained before the vibrated spring is studded. This is done by placing the balance and spring in the movement. The balance bridge, with the stud

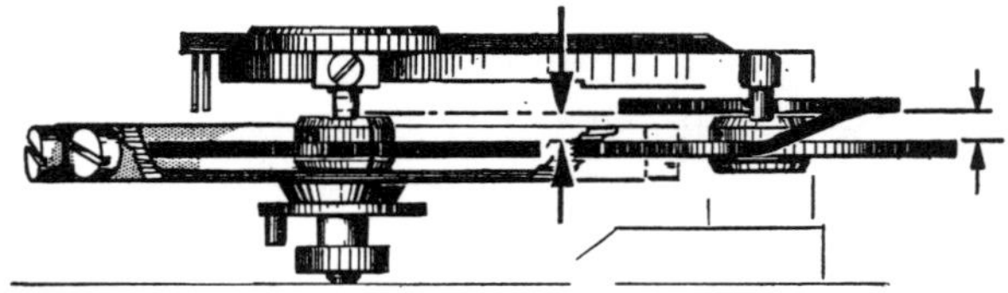

Fig. 97. The height of the overcoil is the distance from the lower level to hole in the stud which is attached to bridge.

fastened so that the top of the stud is flush with the bridge, is also replaced. Then the distance, spring (lower level) to the hairspring hole in the stud, is measured (Fig. 97). This distance is the correct height of the "Breguet."

A small pair of dividers aids greatly in gauging the height. The height of this overcoil must not exceed these requirements, otherwise the spring, when fitted, may appear umbrella-shaped and the lower level of the spring may scrape the balance arms. Generally the height of the Breguet has a relationship to the width of the hairspring, usually between two and three times the width of the spring separates the lower from the upper coils.

One method of raising the overcoil is to bend the spring in a goose-neck or hockey-stick style as shown in Fig. 96. This particular method has certain advantages not enjoyed by the "twist" method. In the "hockey-stick" bend, the spring is raised in ⅛th

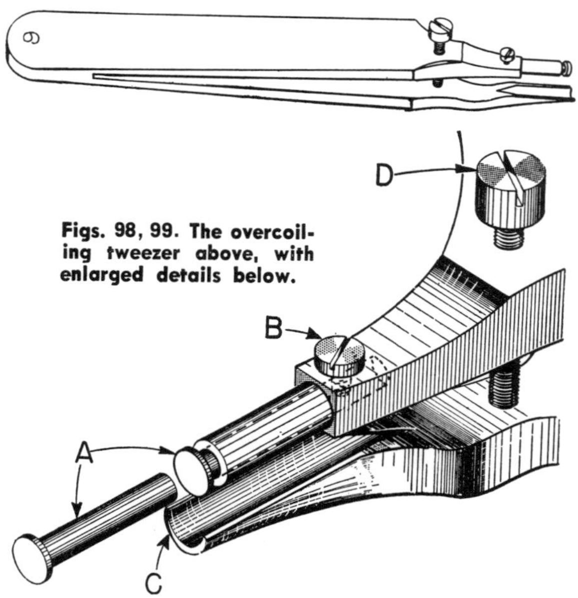

Figs. 98, 99. The overcoiling tweezer above, with enlarged details below.

of a coil, whereas in the "twist" method the spring is required to take a half-coil to achieve its proper height. Thus it cannot be bent inward for a major portion of the riser, otherwise it may scrape the outermost coil it must cross. However, in the "hockey-stick" bend it reaches its new level in ⅛th of a coil, and 1/16th of a coil beyond this it may be bent inward to assume its new curve or shape.

There are a number of methods of making this bend. The one favored by experienced watchmakers and professional vibrators utilizes a special type of tweezers shown in Figures 98 and 99. This tweezer consists of the cylindrical button-topped head (A) which fits into the thick body of the tweezer. This pin may be pulled out or pushed in to accommodate different thicknesses of hairsprings. The set screw (B) fastens it in place after the adjustment has been made. The other tweezer blade (C) is concave in shape, fitting partially around the pin. When the pin and the concave blade are pressed against the edge of the hairspring, it will curve it upwards. The stop screw (D) is adjusted to regulate the angle of the bend, acute or minor.

Fig. 100. Adjusting the tweezer for hairspring thickness.

To use the tweezer, the pin head is brought over the spring so that the outer coil rests between the head and the tweezer edge as shown in Fig. 100. The pin is pushed in so that the spring is barely clamped between the inside of the pin head and the tweezer shoulder, leaving only a minimum of side shake. The set screw is then tightened. The stop screw is adjusted so that only a bend of about 30° is possible.

Before attempting to raise the overcoil on a "live" job it would be advisable to practice on some old hairsprings.

To make the first bend, the spring is assembled on the balance. A regular bench tweezer then holds the balance suspended by the hairspring. This separates the coils enough so that the overcoiling tweezer may reach in and clamp the spring exactly where desired. The tweezer is held so that the pin and head is on top of the spring

Fig. 101. The tweezer grasps the spring at pre-determined spot and first bend upwards is made by squeezing tweezer blades together.

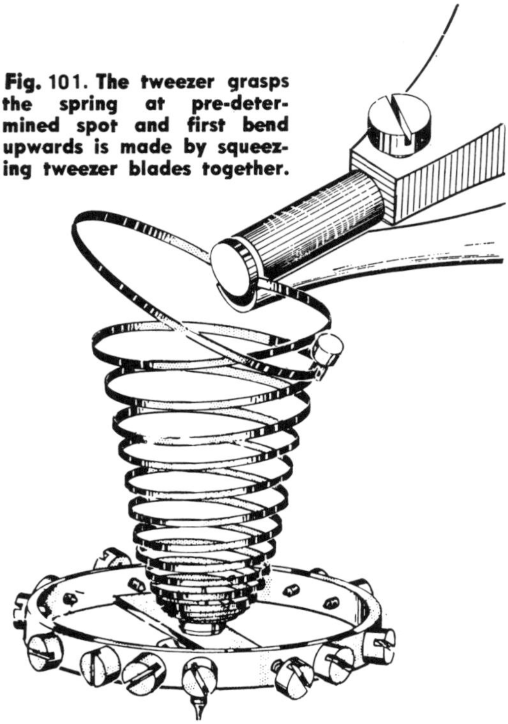

and the concave blade is below the spring. Pressure upon the overcoiling tweezer will cause the spring to bend upwards as shown in Fig. 101.

To level off the hairspring, the tweezer is reversed with the concave blade on top and the cylindrical pin below. The tweezer is

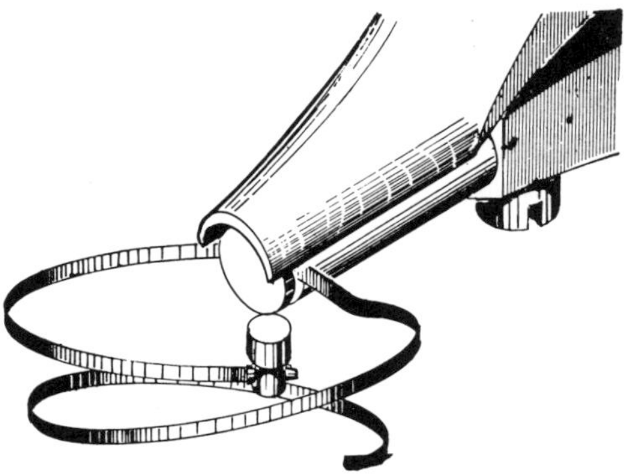

Fig. 102. Leveling off the hairspring after upward bend, tweezer is reversed and applied pressure causes a leveling bend of same degree as the first.

then shifted on the spring ⅛th of a coil from the bend toward the stud. Pressure on the inverted tweezer will cause a levelling bend of exactly the same degree as the first bend, the spring thus becoming levelled. This is shown in Fig. 102.

Since these tweezers are rather expensive and not many watchmakers possess them, there are other methods employed which require less expensive and less elaborate equipment. The elbow or "hockey-stick" bend may be made by placing the hairspring on a block of soft wood, the spot to be bent or raised *crossing* the grain. A stubby, strong and pointed tweezer depresses the spring at the bend spot which causes the sides of the spring adjoining the tweezer to rise. Some watchmakers prefer to use a tapered steel pin, slit very shallow at its tip with a very fine saw. This fits over the

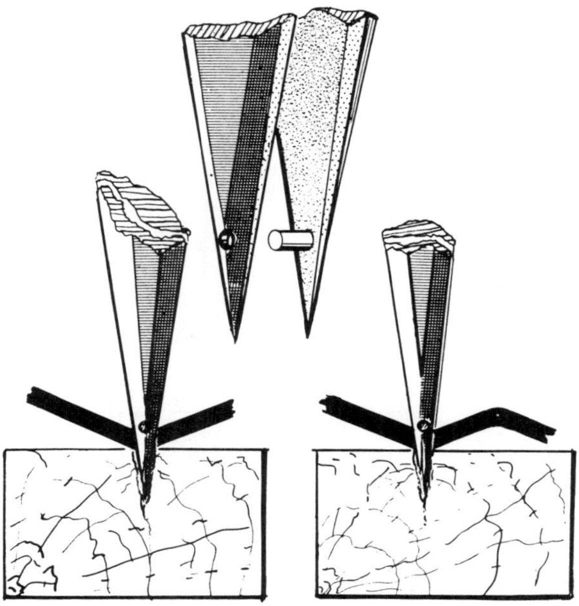

Fig. 103. Tweezer with pin in one blade and hole in opposite blade can be used to make bends, pressing spring into a block of soft wood. Left, first bend; right, the leveling-off bend.

spring and presses it into the soft wood. Others use a pair of tweezers with a pin driven tight inside one of the points and this pin fitting into a hole in its partner blade. This keeps the spring from slipping up into the tweezer while it presses into the wood, Fig. 103.

The bend may be accurately made in this manner by pressing the spring with short light thrusts, bringing up the bend by gradual degrees, thus controlling the angle of the bend.

Placing the spring upside down on the block of wood and repeating this process at a point ⅛th of a coil away will level the spring. Extreme caution must be observed that the second bend is not overdone.

The overcoil may also be raised by twisting the spring. This will cause the highest spot to be diametrically opposite the twist. There are two methods of doing this. The first utilizes two pairs of tweezers held close to one another. One pair acts as the vise while the other twists the spring upwards. The other system also calls for the use of two tweezers but while one acts as the anchor or vise, the other pair grasps the spring on the opposite side of the coil where the highest point of the rise will be. The manipulating tweezer then pulls the spring directly upwards. This will cause the spring to rise gradually.

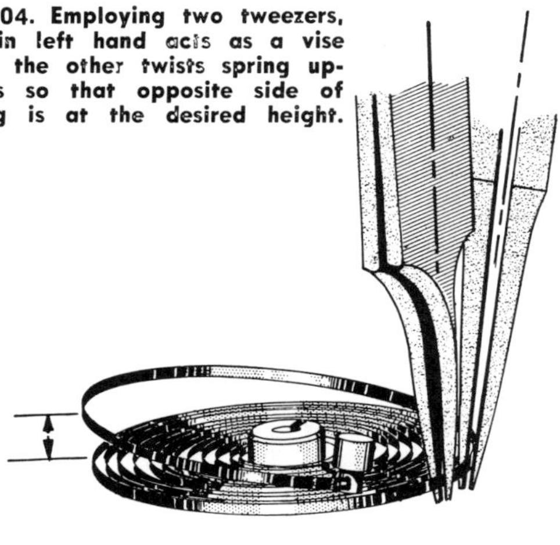

Fig. 104. Employing two tweezers, one in left hand acts as a vise while the other twists spring upwards so that opposite side of spring is at the desired height.

To employ the first method, the hairspring is removed from the balance and placed on a white clear background on the benchplate. The anchoring tweezer is held in the left hand and grasps the spring about ¾th of a coil in from the stud. This tweezer must be held upright and not tilted in any direction. The manipulating tweezer is held in the right hand and grasps the spring about 1/16th of a coil (about two millimeters) away from the first tweezer (towards the stud). The manipulating tweezer in the right hand is then tilted back toward the operator. This will cause the last portion of the spring to tilt upwards. The twisting should be done cautiously so that the highest spot on the twisted spring should not be above the level desired of the overcoil. This is shown in Fig. 104.

The next step is to level the spring so that the remaining portion is parallel with the lower body of the spring. This is accomplished by grasping the hairspring with the anchoring tweezer at the highest spot made by the first bending operation. The manipulating

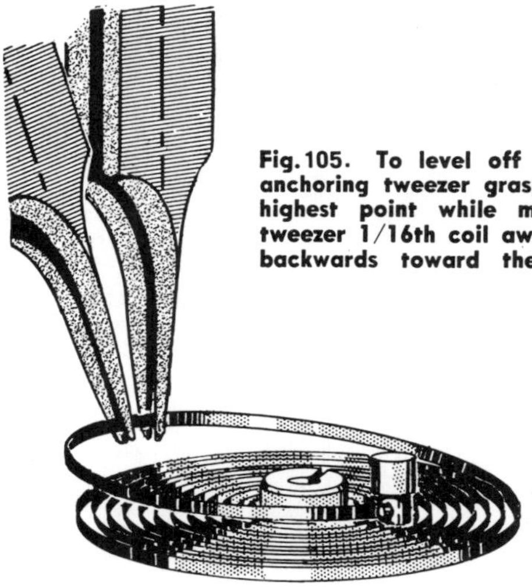

Fig. 105. To level off the spring, anchoring tweezer grasps it at its highest point while manipulating tweezer 1/16th coil away bends it backwards toward the operator.

tweezer grasps the spring about 1/16th of a coil beyond this and the second tweezer is tilted backwards toward the operator. This will bring up the stud end of the spring which until now was pointing downward. This also should be done gradually until the last portion of the spring is parallel with the lower coil. This is shown in Fig. 105.

SHAPING AND CENTERING THE UPPER (OVERCOIL) LEVEL

After the upper coil is made level and parallel with the lower spring, it must be curved and centered. The last portion of the spring must be bent inward and the last coil most always will have to be curved to a smaller circle than the lower coil. The part of the last coil which comes under the influence of the regulator must assume a circular arc. This is required because the regulator also moves in circular arcs and any deviation from such an arc will throw the hairspring off center when the regulator is moved. Most regulators may be moved in an arc of about 30° each side of center and, since the stud is often at right angles (90°) to the regulator, the spring must be circular for at least 120° back from the stud.

To shape this portion of the spring to conform to the above requirements, it is first removed from the balance and the balance

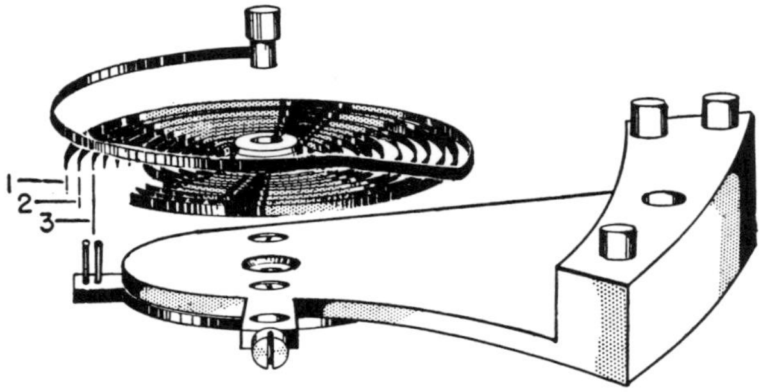

Fig. 106. To aid in selecting the proper radius for the overcoil, the spring is laid over the balance bridge and the coil over the regulator pins as shown above.

bridge is placed on its back. The unattached spring is laid on the bridge with the Breguet on top, in the reverse position it will assume if it were connected to the bridge. The collet center should be directly over the balance jewel center. In this position, the regulator pins, like two close fingers, will be pointing up to one of the coils. Notice at which coil the pins project. Let us say for example that it is the third flat coil from the outside as shown in Fig. 106. Therefore the last 120° of overcoil must be curved in so that it follows the third (lower) coil. Because the lower coil is not circular but part of a spiral, this may only serve as a good but preliminary model or guide.

STARTING THE FIRST CURVE INWARD

In a "hockey-stick" bend, the first curve inward is made at a point just after the spring is leveled or 67° from the first rising bend. In the twist bend, this is not advisable because the spring rises much more gradually and takes a full half-coil before it reaches its maximum height. In the twist bend care must be exercised that the first curve inward does not scrape the first (outer) coil it crosses. Therefore the initial curve should be made about halfway between the first twist and the levelling twist.

Curving the last coil may be done by employing special curving tweezers shown in Fig. 107. These come in sets capable of curving the spring in many radii. This tweezer consists of two blades, one

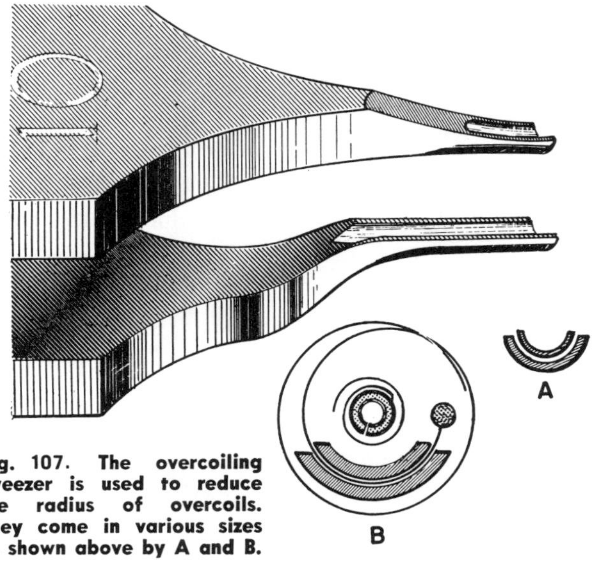

Fig. 107. The overcoiling tweezer is used to reduce the radius of overcoils. They come in various sizes as shown above by A and B.

convex and the other concave. When it is desired to curve the spring to a smaller diameter, the portion of the spring to be shaped is placed between these blades and pressed. The tweezer then is shifted to an adjacent position and the process is repeated until the new arc is obtained.

The spring may be recurved without the use of the overcoiling tweezer by a series of gradual bends which must not kink the spring. The spring may also be recurved by skillfully using an ordinary bench tweezer, one tweezer grasping the spring at one

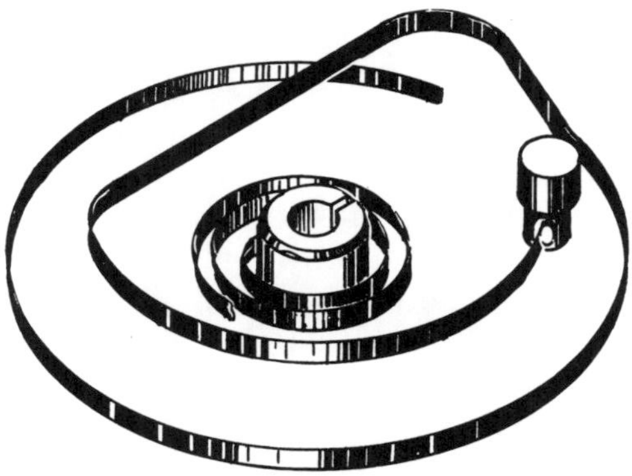

Fig. 108. A type of Breguet overcoil where the last coil is brought straight across spring to avoid a too-close center wheel.

point and the other, let us say, halfway across. Bringing both tweezers towards one another will buckle or bend the spring. Doing this along the spring at skillfully selected spots will, in effect, reduce the radius of the spring.

When the overcoil is curved, it is not yet attached to the balance, but placed correct side down, stud in the bridge. The bridge with the attached hairspring is then replaced in the movement. The regulator is moved along its full compass. If this throws the

Fig. 109. This shows how close center wheel makes necessary a bend in the spring as shown at right.

spring off center, the part of the spring within the regulator's influence must be bent or curved until the spring is not disturbed when the regulator is moved.

The spring is then removed from the bridge, attached to the balance and then reattached to the bridge. When the connected unit is in the watch and going, the spring may be touched up so that it is level and "breathes" evenly.

In some watches the center wheel is so close that it may interfere with the overcoil. In this case, the overcoil may be shaped as shown in Fig. 108 so that there is no contact with the wheel. Sometimes an overcoil like the one shown in the illustration will be seen that has no danger of contacting the center wheel. Curves like these may be designed for specific reasons best suited to the balance unit for isochronal performance, although when a case is met with such as shown in Fig. 109, no other choice may be left.

Final checking points to be observed are, one, to make certain that the levels are flat and parallel; two, that the lower coil does not scrape the stud or the balance arms; three, that the overcoil does not touch the center wheel or scrape the balance bridge; four, that the riser does not scrape the outside coil as it curves inward, and five, that the coils "breathe" evenly all around.

BOOK II

CHAPTER I

REPLACING REGULATOR PINS

ONE of the most frequent causes of poor timing in watches may be directly traced to the poor condition of the regulator pins. A regulator with a broken pin may cause a loss up to 20 minutes a day. Bent, worn or burred pins also cause inconsistencies in timekeeping. Although some new watches are appearing without regulators, these are in the extreme minority so that a knowledge of replacing these pins is still an essential part of the watchmaker's skill and trade.

Replacing a regulator pin may become necessary because of breakage or because it may have become marred, irreparably bent, or kinked, as shown in Figure I. Bent regulator pins cause the hairspring to waver so that the vibrations are irregular.

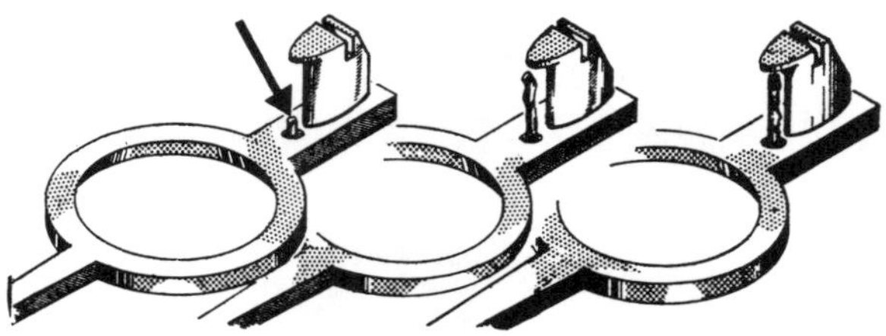

Fig. 1. Broken, bent or marred pins must be replaced.

To remove a broken regulator pin, the regulator is removed from the balance bridge and the regulator is placed topside-down on a steel bench block. With a sharp well-fitting screwdriver, twist the regulator key completely around so that it appears as shown in Figure 2. Then adjust the regulator so that the stump of the broken pin is directly over a suitably sized hole in the steel block. A sharpened steel needle, set into an appropriate holder, is then placed on the stump of the broken pin (Figure 3) and downward

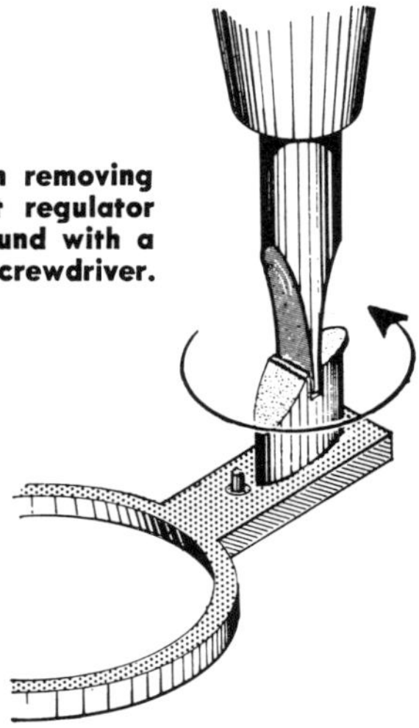

Fig. 2. First step in removing the pin is to twist regulator key completely around with a sharp, well-fitting screwdriver.

pressure is applied. Sometimes the original pin is lodged rather securely in its hole in the regulator, and some force may be necessary to dislodge it.

Regulator pins can be bought, but sometimes, few of these will

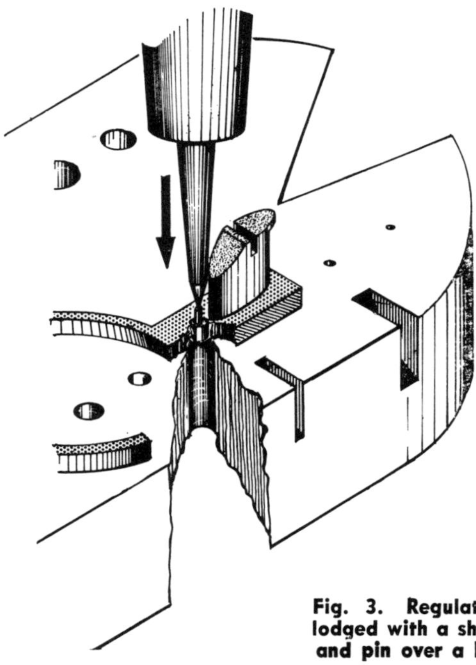

Fig. 3. Regulator pin is dislodged with a sharpened needle and pin over a hole in a block.

fit a particular regulator owing to the great variance of these holes. Therefore, it is important to know how to make a regulator pin, should the occasion arise.

To an apprentice, making a new regulator pin is not as simple as it seems. It is not enough to insert a tapered pin to fill up the hole. Regulator pins should present a cylindrical and perpendicular appearance. A tapered surface will cause a slight shimmy-

Fig. 4. A tapered pin will cause a slight shimmying of the hairspring as only the bottom edge of spring touches regulator pin.

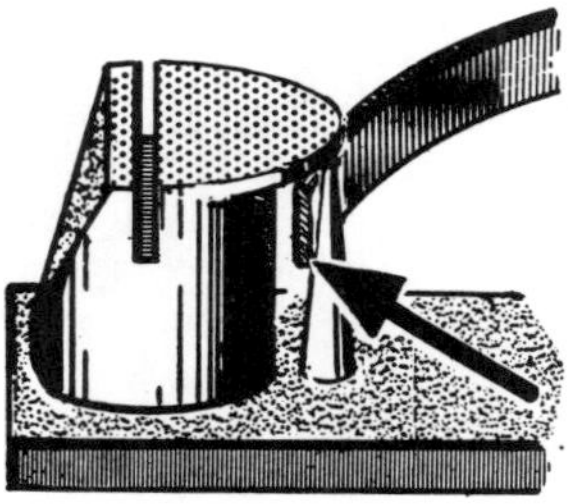

ing of the hairspring. Figure 4 shows how only the bottom edge of the hairspring touches a tapered surface (pin). Here the bottom edge of the spring touches the slanted surface of the outside pin and as it continues to press against this pin, the incidence of these angles forces the spring upward. When the spring expands and is forced against the other surface if it is tapered, the opposite effect will take place, thus the hairspring will have an alternately upward and downward shimmying, all owing to improperly shaped pins.

In order for the pin to become secure in the regulator, the section of the pin that will remain in the pin hole should be slightly tapered, but the part of the pin regulating the hairspring should be cylindrical as shown in Figure 5.

Too thick a pin does not leave sufficient space for the hairspring and a slight amount of necessary freedom. This is shown in Figure 6. If the spring is tight between the pin and the regulator key, any movement of the regulator to bring the balance to time will bend the hairspring as shown in Figure 7.

Regulator pins may be made from hardened brass, nickel, or wire. Straight brass pins are ideally suited for conversion because these are hardened and are of sufficient thickness to permit handling and reduction in diameter without too much waste. The first operation is tapering the wire until it enters the pin hole in the regulator and extends through it as shown in Figure 8. Tapering the pin can be done in the lathe or in the pinvise, skillfully manipulated. When made by hand, the brass taper pin is held in the pinvise and the pin is rested in a shallow groove of a wooden block held in the benchvise.

Ready-made taper pins are about ¼ in. (6.35 mm.) long. These have a taper that gauges about .60 mm. at one end (thick end) and tapers down to about .25 mm. at the other (thin end). This insures

Fig. 5. Only bottom part of the pin to remain in pin hole should be tapered to securely fasten it.

a taper of the correct angle—one that will easily be secured and not work loose.

When the pin is made in the lathe, a fine flat file could be used

Fig. 6. Too thick a pin does not allow the space necessary for freedom of the hairspring.

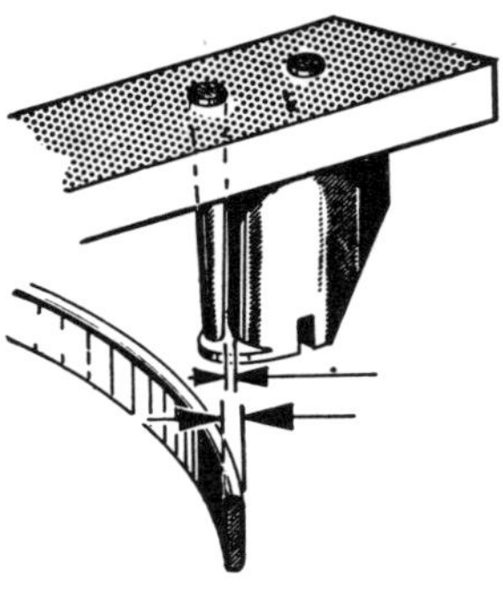

to bring the taper to shape. The file is held underneath the wire so that the work may be seen. In this case, the direction in which the lathe generally turns must be reversed so that the file can cut.

Fig. 7. Too little space between the pin and regulator key will bend the hairspring when the regulator is moved.

Care must be exercised not to press too hard up against the taper as this will bend it and make it appear as shown in Figure 9.

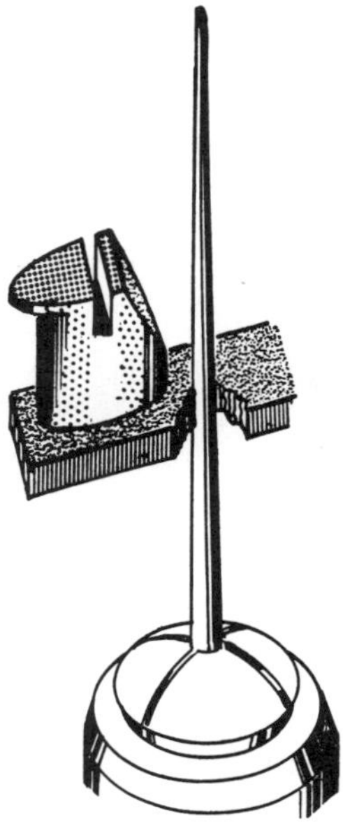

Fig. 8. First operation in making the pin is tapering it until it enters the pin hole as shown.

Some watchmakers prefer to use a Jacot runner or a steel block with groove to support the pin while the taper is being formed. In

Fig 9. Too much pressure against pin with the file while turning in the lathe will cause result as shown.

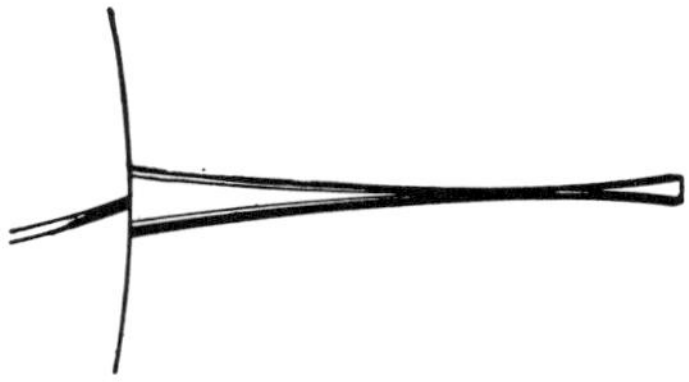

such a case the file is run across the top of the pin instead of under it.

After the first taper is formed, the regulator is fitted over the taper as shown in Figure 10 and pushed as far as it will go, twisting it on with a little pressure. Then the regulator is twisted off again. The place on the taper pin that was tightest in the regulator

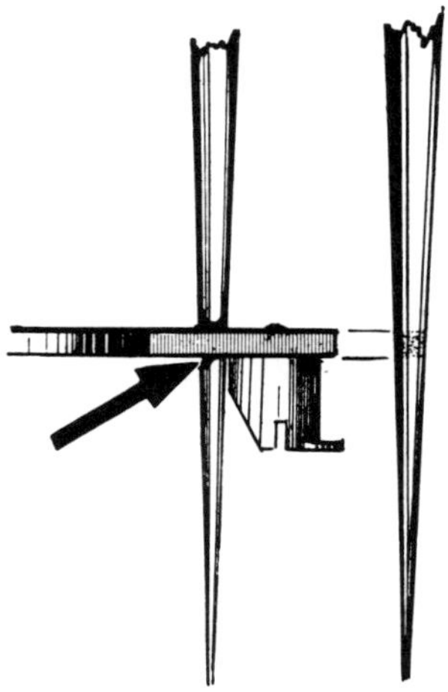

Fig. 10-11. The regulator is twisted on taper as tight as possible. This spot will be indicated by burnish mark and from this point onward, the pin is made cylindrical.

pin hole will show a burnish mark (Figure 11). This point is noted because from that mark outward the pin must be reduced in thickness so that it becomes cylindrical, or a file mark is made at the spot the taper pin emerges from the pin-hole in the regulator (shown by the arrow in Figure 10). This too will indicate where the tapered section of the pin is to be converted to a cylindrical shape.

Before this operation is started, this section is shortened to its proper length. The length of a regulator pin must be adjusted so that it will permit the toe of the regulator key to be closed under it (Figure 12). It should not be too short so that the hairspring can escape from it (Figure 13). The length of the regulator pin is adjusted at this time because it is still thick enough to have its end filed flat, which could not be done if it were thin. Furthermore if it were left longer it would be difficult to make it cylindrical and as thin as required.

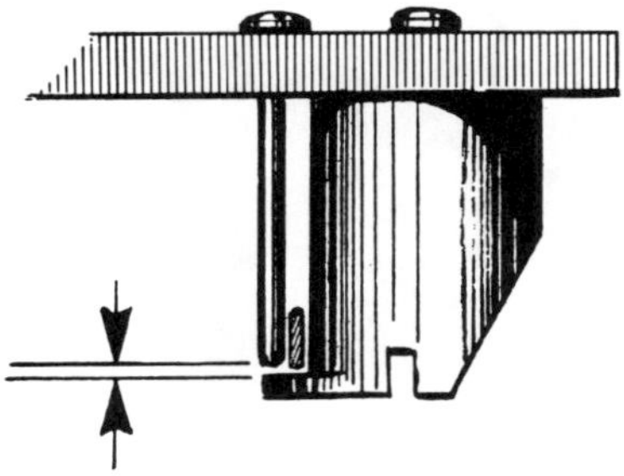

Fig. 12. Before tapered pin is converted to cylindrical shape it must be cut to proper length.

To determine the correct length, the regulator key is turned partially back again, and the pin is cut off flush with the bottom surface of the key. Then, the file works across the front tip of the pin until it is about .05 mm. above the top of the toe of the key,

Fig. 13. Regulator pin must not be cut off too short so as to allow the hairspring to escape from under it.

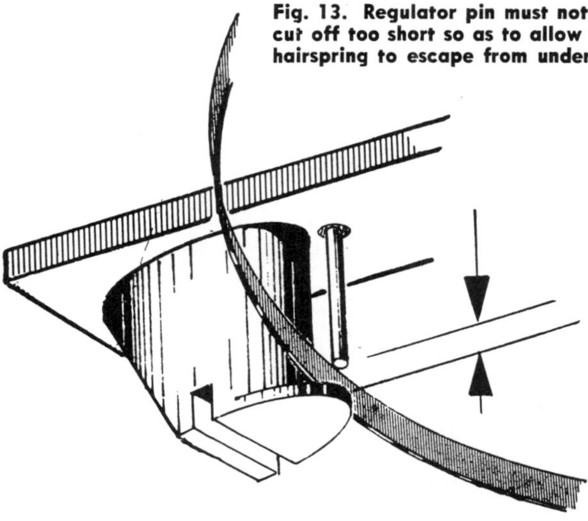

leaving this much freedom between the top of the toe of the key and the end of the pin.

The cylindrical portion of the regulator pin when finished, should gauge about .10 mm. However, the filing should be stopped when the pin reaches .15 mm. Here, a burnisher or a fine jasper stone is used to finish the pin so that when it is reduced to its final thickness

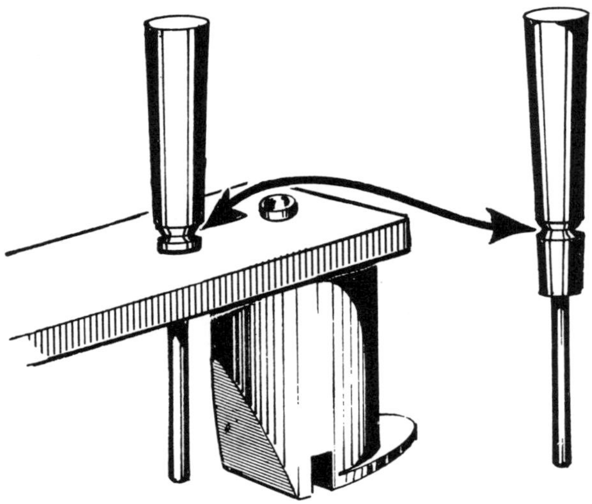

Fig. 14. Groove is cut into section of the taper just above point it enters hole to facilitate breaking this portion off.

it will have a fine finish and be free from file marks. When this is done, it should look like Figure 14. To facilitate breaking the pin after it is turned to dimensions, a groove is cut into the section of the taper which is just above the point where it *enters* the pin hole in the regulator, shown by the arrow in Figure 14.

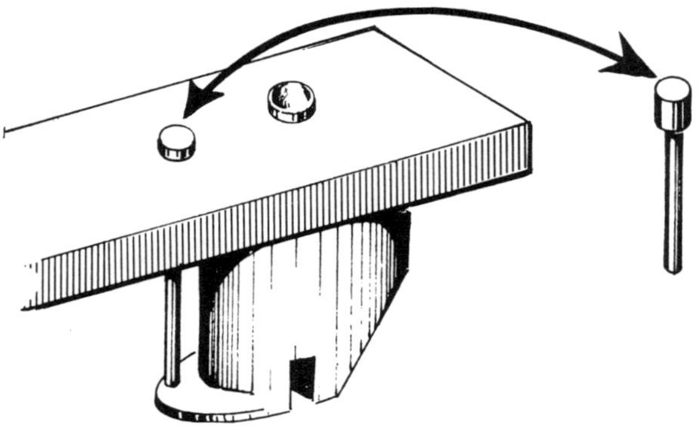

Fig. 15. After breaking off unwanted portion above the groove, top of pin is finished off with file, burnisher or jasper stone.

After this operation, the regulator is twisted on the pin as tight as it will go. A slight bending twist will break the pin at the groove (arrow—Figure 14). The top of the pin is then finished off with a file, burnisher or jasper stone, taking care not to scratch the polished steel top of the regulator. The finished job should look like Figure 15.

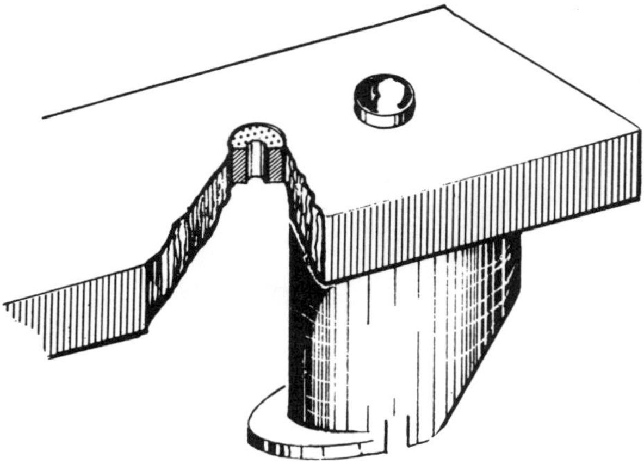

Fig. 16. Preparing old pin for repair, top is filed flush with the regulator and hole is drilled in center of stump of old pin.

Another method of repairing a broken regulator pin is to file the stump of the broken pin flush with the underside of the regulator. Then a very fine pivot drill, set into the nib of a bench screwdriver is made to drill a hole through the center of the stump, as shown in Figure 16.

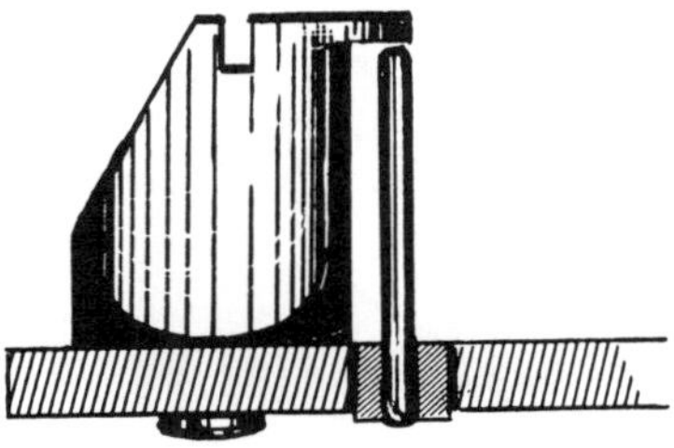

Fig. 17. A fine piece of wire is inserted tightly in drilled hole in stump and its length is adjusted to fit regulator key.

A fine piece of wire is then prepared which will fit tightly into this hole, and the length is adjusted to suit the regulator key. This finished job should appear as shown in Figure 17.

Replacing pins for a Breguet regulator is easier than furnishing those for a flat hairspring. The pins may be thicker since there are no adjacent coils that can contact them. Generally, the space required between the two pins of a Breguet regulator is not as great as between the pin and key of a "flat" regulator. However, these pins must be perpendicular, parallel to one another and present a smooth appearance.

The pins of a Breguet regulator must not be too long or they might protrude down into the lower level of coils. However, they

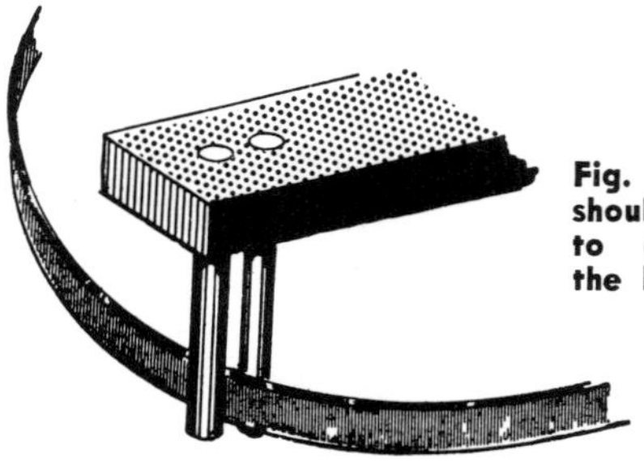

Fig. 18. Regulator pins should be only long enough to extend slightly below the bottom of the overcoil.

must be long enough to contain the overcoil. This is shown in Figure 18, where the regulator pins straddle the overcoil but are long enough to extend only slightly below the bottom of the overcoil.

The very ends of Breguet regulator pins must not be burred because, as shown in Figure 19, it will be difficult to place the overcoil in between the pins. Furthermore, burred pins present the danger of the lower coils getting caught on the shelf of the burr should the watch receive a bump that would cause the lower coils to be jarred upwards.

To remove any burrs from the inside ends of the pins, a fine screwhead file is used. To prevent the lower coils from getting caught on the burrs and to provide additional clearance in such a case, the outside ends of the pins should be filed and bevelled as shown in Figure 20. This should be done preferably with a fine Arkansas oilstone slip as a coarse one may raise burrs itself.

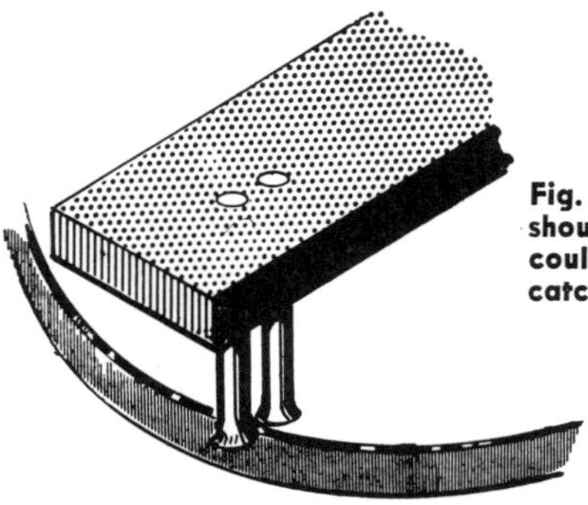

Fig. 19. Bottom of the pins should not be burred as this could cause lower coils to catch if watch receives jar.

The space between Breguet regulator pins should be close enough to brace the hairspring without any noticeable light between the pins and the hairspring but yet not so tight that it binds the spring so that any movement of the regulator index causes the spring to buckle as was shown in Figure 7.

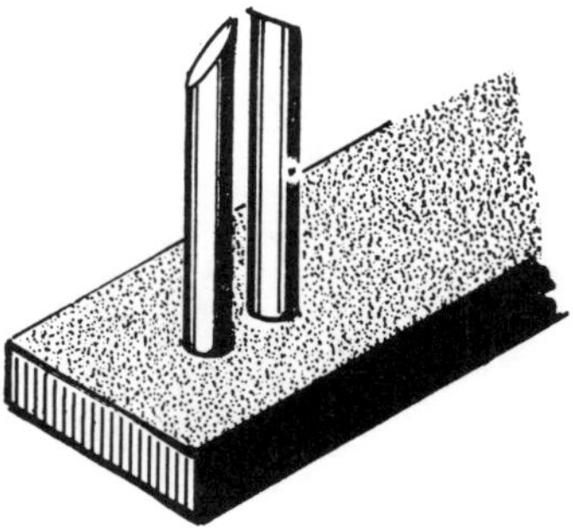

Fig. 20. Outside ends of pins are bevelled as shown to prevent catching of lower coils.

When the pin holes in the regulator are drilled far apart so that the pins appear as shown in Figure 21, these then must be brought closer together to comply with the requirements stated in the previous paragraph. However, they must not be bent as shown in Figure 22 as this, in turn, will cause the error that was illustrated in Figure 4. When the pins are planted far apart, they may be brought closer together by bending these pins so that they appear as shown in Figure 23. Of course this operation shortens the pins so that this must be taken into consideration when the pins are made.

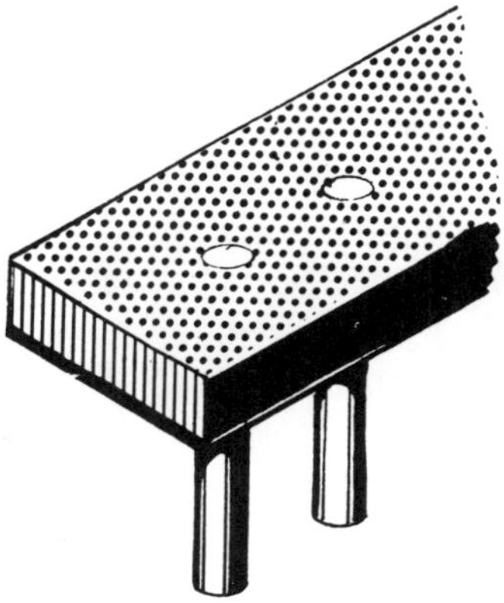

Fig. 21. Pins placed too far apart must be brought together to brace the spring.

To bend the pins as in Figure 23, a special pair of tweezers is recommended. These tweezers are shown in Figures 24 and 25 and are very useful for adjusting Breguet regulator pins. Figure 24 shows the tweezer into which, a groove (B) is cut in each point. These grooves fit partially around the regulator pins. A very thin strip of mainspring wire "A" is riveted to the lower inside of one of the tweezer blades at "C."

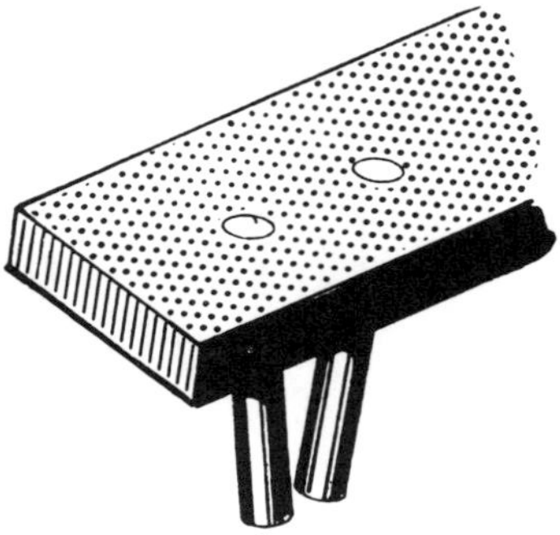

Fig. 22. Pins bent together as shown will not support a hairspring along its width and will cause it to shimmy.

In operation, the tweezer is placed over the regulator pins so that the spring "A" (Fig. 25) is between the regulator pins, and the grooves "B" straddle the pins. Pressing the tweezer points

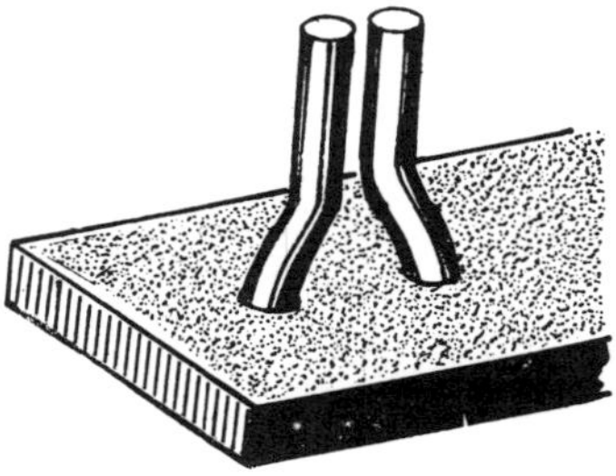

Fig. 23. Pins should be bent together like this so as to give full support to spring.

together will bend the pins closer together, the amount depending upon the pressure applied. The pins will be upright and parallel. The spring tongue "A" keeps the pins from closing altogether or from scratching each other. These tweezers can be made from an old pair of general work tweezers. The grooves are best made by drilling a thin hole downward into the points while these are tightly clamped together, each point containing half a (deep) hole. The tongue "A" is made from a strip of thin, narrow bracelet mainspring. This spring should be no thicker than .07 mm. Other details are clearly shown to guide anyone interested in making up a pair.

When regulator pins are bent as shown in Figure 26, these may be straightened with the regulator pin-tweezer by inserting the tongue blade between the bent pins and the grooves around the pins and twisting the pins to an upright position. Pressure upon the tweezer will rectify any small errors still existing.

When regulator pins must be spread apart, this may be done with a needle, ground flat, sharp and very gradually tapered. The sides

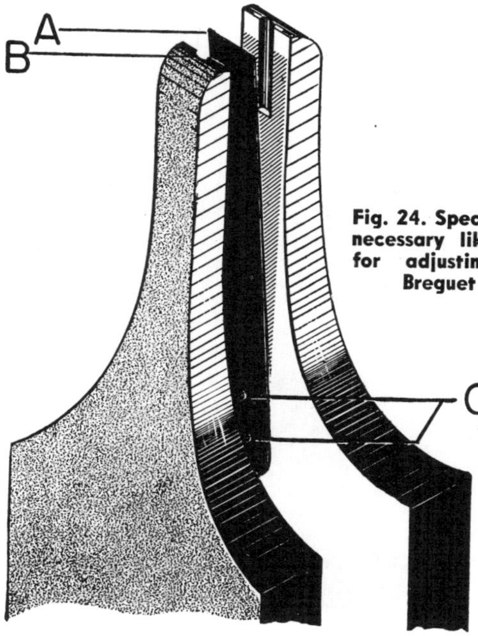

Fig. 24. Special tweezers are necessary like these shown for adjusting pins in a Breguet regulator.

should be rounded off and the flat tapered surfaces of this blade should be highly polished to prevent scratching the inside of the

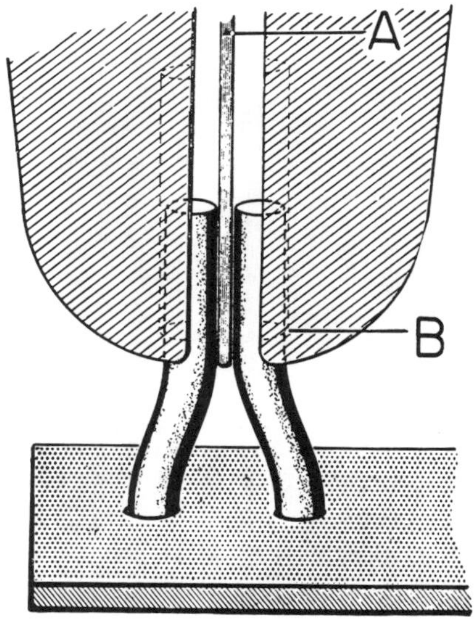

Fig. 25. Showing how the tweezer is placed over the regulator pins with spring (A) between pins.

pins. This miniature screwdriver-like blade is inserted in between the pins in a light wiggling motion, making sure that the pins are not being bent *away* during this operation.

Figure 27 shows a type of regulator used with Breguet hairsprings. This has the key or "boot" of a regulator used with a flat hairspring and has two straight pins which contain the Breguet overcoil. The regulator is, of course, a Breguet type of the conventional two pins but with the key added to keep the overcoil from

Fig. 26. Regulator pins bent as shown may be straightened with the special tweezers.

escaping from the pins should the watch be jarred. Repairs to this type of regulator are made according to the combined instructions given for both flat and Breguet regulators.

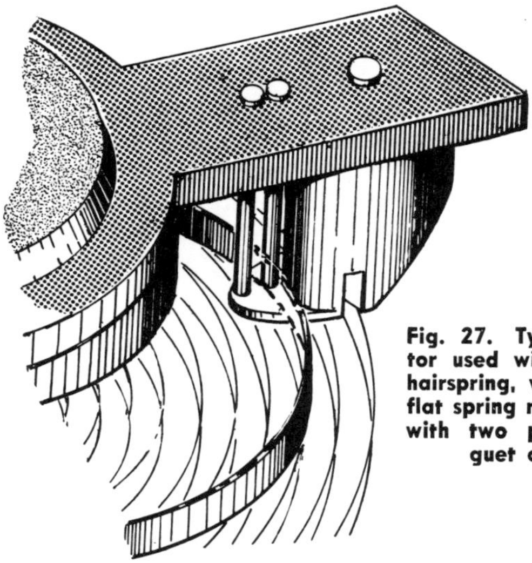

Fig. 27. Type of regulator used with a Breguet hairspring, with key of a flat spring regulator used with two pins for Breguet overcoil.

CHAPTER II

Part I

VARIOUS TYPES OF SCREWS AND HOW TO MAKE THEM

SIMPLE as they are individually, the varied assortment of screws used in a watch are one of the most complex facets of the watchmakers' trade. In this chapter, these various screws and some of their differences are explained.

One of the most paradoxical branches of watchmaking centers around watch screws. No universal agreement exists on thread size, pitch, design or thread direction.

In one watch, one may find as many as four different kinds of bridge screws, ratchet and crown wheel screws with different thread direction, cap and jewel screws of various thicknesses and lengths, some with polished heads, others with unpolished heads but polished thread ends and other differences. A listing of these screws and some of the differences found in watches of ordinary construction might follow in this order:

Bridge screws: Long, short, thin heads and thin threads, regular threads but short heads.

Jewel screws: Unfinished heads, polished heads, polished thread ends (only), very short threaded sections.

Case screws: Long full heads, long half ("D") heads, short full heads, short half-heads.

Crown wheel screws: Left thread, right thread, shoulder screws.

Ratchet wheel screws (also called barrel arbor screws): right and left threads.

Set lever screws: Hip-shoulder screws; long-head shoulder screw.

Stud screw: Wide, short head, long thin screw.

Click screws: Wide, short head, thin thread; shoulder screws, combination click and bridge screw.

Pallet bridge screw: Regulator spring screw-micrometer (regulator) screw, setting bridge screw.

Dial screws: Hip, side and top-dial screws.

To be able to recognize their differences and the reasons for these dis-similarities as well as the ability to make these screws is an important facet of the watchmakers' trade.

One should realize that a varied assortment of screws in one watch raises the cost of production considerably. This is due to the many set-ups required for production as well as the problem of assembling the unit. Therefore, it must be reasoned that each screw that differs from one another in the same category in one watch, is made so purposely.

The design of a screw is primarily utilitarian; it must serve a specific purpose. After this, it may be made to improve the appearance of the watch. This may take the form of a polish, bevel, or color. Sometimes a screw is made blue or wine color to make it stand out from others of similar size, and its individual color serves as a reminder of its unique purpose. Such a screw in a pocket watch with a negative setting may indicate that this one may be tightened to bind the set parts in the winding position in preparation for timing the movement out of the case.

BRIDGE SCREWS

Bridge or plate screws are made as long and as heavy as space will permit and in keeping within the reasonable limits of security desired. This also must consider the assembler's method of manipulation and his ability to tighten or efficiently loosen the screw. For instance, space may permit a much heavier screw but this may require too much strength to tighten or loosen it without damage to the plates. Towards this end, the width and height of the head, the length, thickness and pitch of the thread must be considered.

When a bridge screw is shorter or thinner than the others in a watch, it is because it is made to work where there is less room. For instance, a barrel bridge generally has three bridge screws to secure it to the pillar or main plate, two of these being at opposite ends of the bridge and one somewhere between these two. The threads reach almost through the lower plate, close to the dial. However, one of these comes out directly behind one of the setting parts, generally the set lever. To avoid contact with this thin piece, its threaded section is made shorter than the others. This screw hole is easily identified by two simple methods. One is to look into the screw hole. If light can be seen clear through it, this may take a screw of regular length, if not, it may require a shorter screw. Also, this screw hole is close to the detent screw. This is shown in Figure 1. To replace this with a screw of ordinary thread length may cause its end to push the set lever outward and break the lever off its screw.

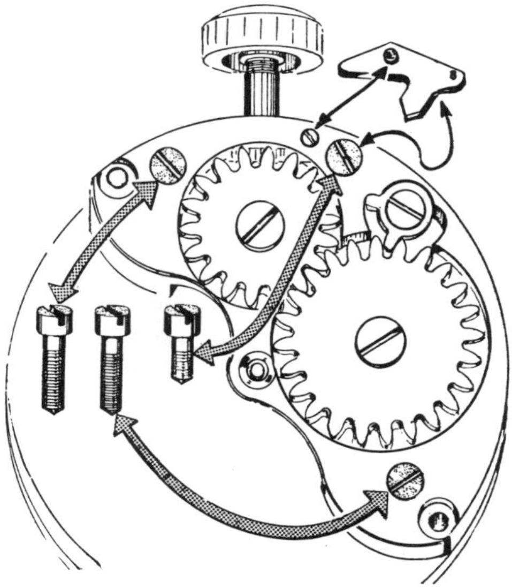

Fig. 1. Of the three screws generally used in a barrel bridge, one is shorter to avoid contact with the setting parts, generally the setting lever.

Watch designers are aware that modern cases are made to make the watch appear thinner. These cases have such close tolerances that the inside of the case backs may actually make contact with the movement and thus reduce or eliminate the endshakes of one

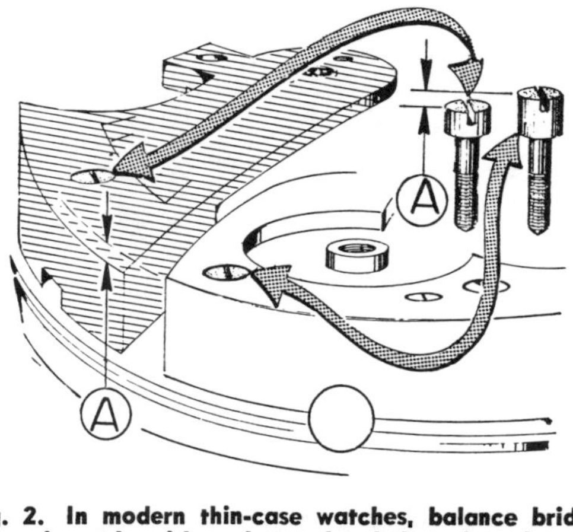

Fig. 2. In modern thin-case watches, balance bridge screw is made with a shorter head than the others to guard against possibility of case pressing on bridge.

or more parts. To guard against the possibility that the balance bridge may be pressed by the case back, it is often made a little lower than the other bridges. Therefore, this bridge screw is sometimes made with a shorter head than the others. This is shown in Figure 2. While a screw of standard size may work, its head would stick up above the surface of the bridge. It will also require that its own screw serve elsewhere with its head below the surface of a bridge with the possible scoring of the screw-hole.

When a bridge screw differs from the others in thickness, it is because the space around its hole on the bridge is crowded. This may be in an escape wheel bridge where the wheel is planted close to the edge of the movement, leaving little room for the post of the bridge. This screw and its head must be thin so that it remains within the narrow confines of this bridge. Another spot often requiring a thinner bridge screw is the pier of the train bridge that supports the center wheel. The section of this bridge which is wedged in between the balance cock and the barrel bridge, sometimes is so narrow that there is little room for anything but a thin

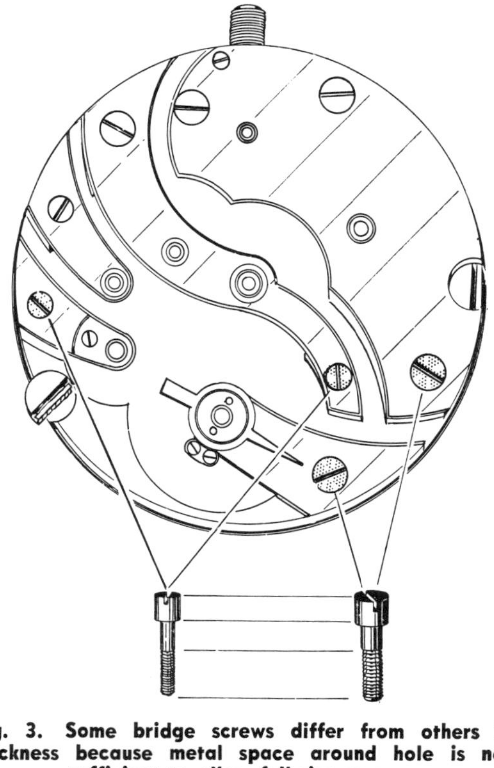

Fig. 3. Some bridge screws differ from others in thickness because metal space around hole is not sufficient to allow full-size screws.

screw. This situation exists especially in very fine watches. This is shown in Figure 3.

To facilitate the selection of the correct screw for reassembly of the movement, some watchmakers place each screw into a piece of flat pithwood in its relative position in the movement. This method is illustrated in Levins' "Practical Benchwork for Horologists." Material supply houses sell hollow metal blocks with the shape of a movement etched on its surface. These have numerous holes drilled

Fig. 4. To facilitate the selection of the correct screws in reassembling a movement, a metal block with shape of movement etched on it is used.

into them into which is placed the screws as they are removed from the movement, and set into the screw block in the relative position they occupied while in the movement. Such a block is shown in Figure 4. However, most watchmakers today prefer to clean the screws together with other parts, placing them in the mesh basket of the cleaning machine. This requires identification and separation later.

A simple method used to determine the correct screw for the bridge holes is to select what appears to be the screw for a particular bridge hole and taking the screw and turning it upside down, place its head into the hole, with the threaded section of the screw pointing upwards. Then notice whether the head is flush with the top of the bridge and whether its width fills up the diameter of the

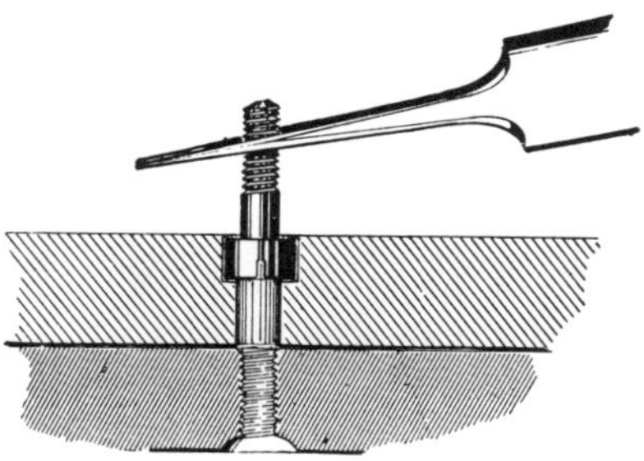

Fig. 5. Turning screw upside down in a hole is simple method used to determine correct screw for bridge hole.

hole. In this way, it is possible to avoid having to thread the screw in all the way with the chance of stripping the threads and marring the plate around the hole. This method is illustrated in Figure 5.

Where a screw must necessarily be a short one, its thread or gripping surface must contact sufficient area to compensate for its lack of depth. Furthermore, if a part or bridge is comparatively long and only one screw can be used, then its head should be wide enough to supply the necessary coverage and pressure to keep this bridge snug against the plate. The pallet bridge is just such a

Fig. 6. Single screw holding pallet bridge must be of sufficient thickness and with a wide head to provide the necessary pressure to keep the bridge snug against the plate.

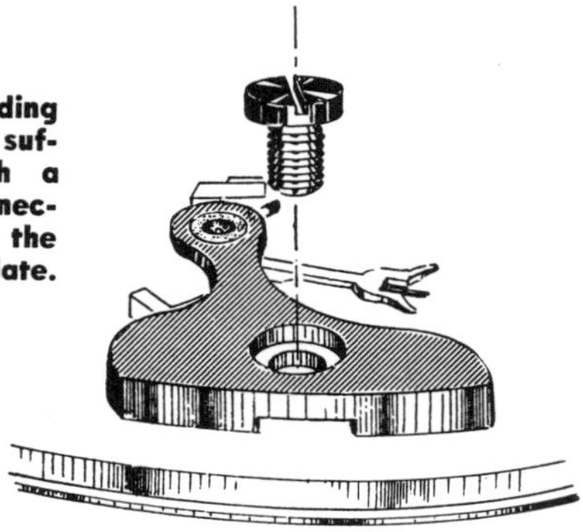

part. The plate below it is very thin at this point, affording height for only a few threads. It is held down by only one screw. Therefore that screw must be comparatively thick and with a wider head than a bridge screw. Because the balance moves above it, its head must be a very short one. This is shown in Figure 6.

CROWN AND RATCHET SCREWS

Very often the ratchet and crown wheel screws look alike, causing them to be interchanged. Sometimes, however, one or the other is left threaded. When these are both threaded in the same direction, they may be replaced incorrectly. If these two screws are placed side by side, it will be noticed that one is shorter than the other. The shorter is the crown wheel screw. The ratchet screw goes into the barrel arbor which is drilled and threaded almost to its full length. This permits the screw to be fairly long. The length of the crown wheel screw is limited because the clutch wheel is directly below it as shown in Figure 7B. Should these screws be interchanged the longer ratchet screw might butt against this wheel and prevent it from shifting into the setting position. This might also cause the threads in the bridge to strip or warp the crown wheel boss.

Crown wheel and ratchet screws often come in left (counter-clockwise) threads. One reason for this difference is that in most watches the crown wheel is turned in a counterclockwise or left direction. In these watches, the screw heads hold down the crown

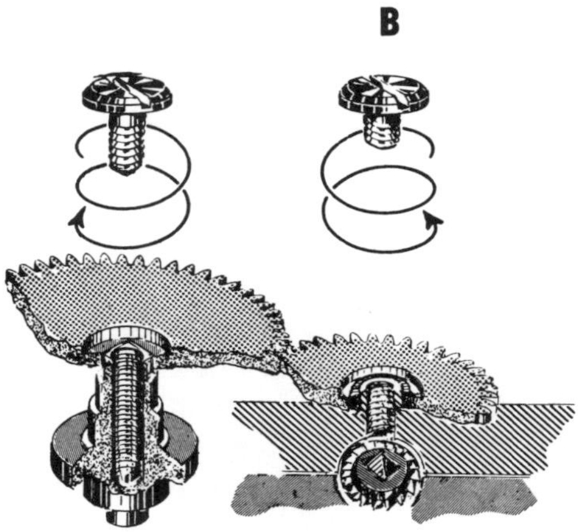

Fig. 7. Crown wheel and ratchet screws are similar in appearance but length of former (B) is limited to avoid contact with the clutch wheel below it.

wheel. Since the crown wheel turns in the "left" direction, its continued rubbing and contact with the screw may eventually loosen it. Therefore, this screw is made to tighten with this counter-clockwise contact.

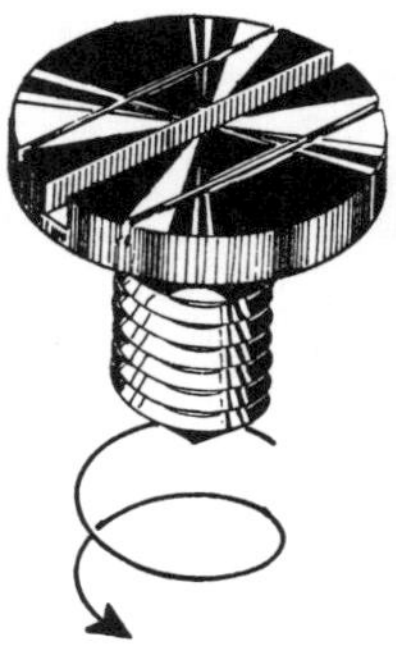

Fig. 8. Any unusual design on the head of a screw indicates caution in using as some manufacturers mark the heads of screws to show the direction of the threads.

If a crown wheel is held down with a wide washer—one that does not turn—a screw of conventional thread direction (right) may be used. In these cases, the washer is kept from turning by a small off-center stud rising up from the movement and fitting into a recess or hole in the washer.

Ratchet screws often have left threads too, but for another reason. The ratchet screw holds down the ratchet wheel and moves only when the watch is being wound. In these watches, the ratchet or crown wheel screws are the highest parts of the movement. Should the case back be soft and touch the ratchet screw during winding, the ratchet wheel would turn but the screw pressed by the case back would remain stationary and thus unscrew itself. When watches come in for repair and the cause is a loose ratchet wheel or screw, most often the case back will bear the impression of this screw.

In watches with thin solid gold cases, this fault often occurs. Watch movements in water-resist cases also bear this tendency after repair. Originally, the cases are designed to provide clearance for the back of the movement but sometimes after repair, its rubber, plastic or metal gasket is not replaced. This not only nullifies its water resisting properties but permits the case back to be screwed or fastened down further and then contact with the ratchet or crown screw is made. For these reasons, in order to prevent the screw from becoming loose, it is made left-threaded so that should contact be made, it would have a tendency to become tight instead.

In some modern, sweep-second watches, the barrel bridge is lower than the train bridge. In such a case, the ratchet screw may have the conventional right threaded screw.

When a watch is received where the direction of these screws are in doubt, it is necessary to *feel* the screw's resistance to the screwdriver in order to loosen it. This is done by trying to loosen the screw in the conventional counterclockwise direction, using only a moderate amount of force. If this fails, then the reverse direction is tried, using just a little more power. Most experienced watchmakers using this method are able to *feel* the direction in which the screw may be loosened.

Some manufacturers make an effort to inform the watchmaker of the presence of a left-threaded screw. Waltham and some Swiss companies slot their screws as in Figure 8. Others blue them. Some watches simply imprint a directional arrow around the screw-hole. At any rate, any unusual design on a screwhead is an indication that caution should be exercised in its manipulation.

BEVEL-HEADED SCREWS

Most screws are designed so that their heads are flush with the tops of the objects they bind. This conserves needed space and

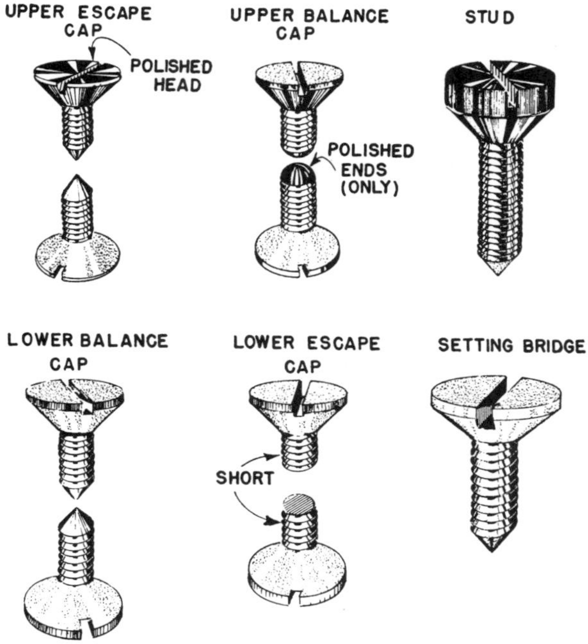

Fig. 9. Types of bevel-headed screws and their uses.

height. When one thin object must be secured to another, there is no room for the countersink to accommodate a flat head or fillister

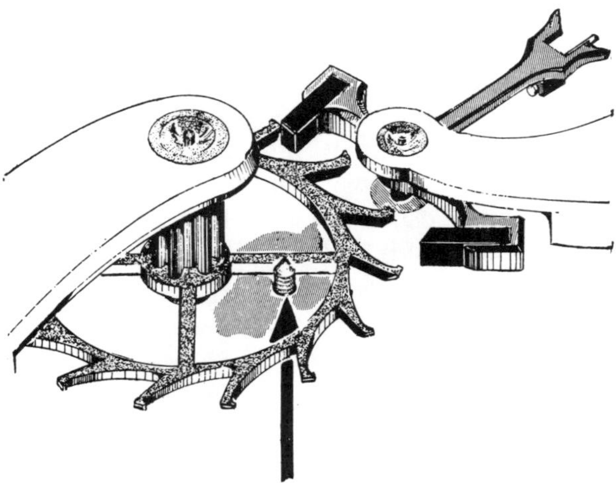

Fig. 10. With two lower cap jewel screws similar except for length, shorter is used for lower escape cap jewel as the longer one would come up between spokes of the escape wheel as shown above.

screw. This obstacle is overcome by the use of a bevel-head screw. In the family of bevel-head screws used in a watch there are setting bridge screws, upper cap jewel screws, upper escapement cap jewel screws, lower cap jewel screws and top-dial screws. These are the smallest screws and it might be difficult to tell them apart. A simple clue to their identity is their polish or lack of it. Let us say there are four bevel head screws of identical measurements. However, one is entirely unpolished, two are polished only at their thread-ends and one has a polished head. Bearing in mind that a polished surface is meant to please and meet the eye, it should be easy to recognize them.

The screws that are polished on their thread-ends only while their heads remain unfinished, are the upper (balance) cap jewel screws, since these are fastened to the cap jewels from underneath the balance bridge. The heads of these screws are not seen so the heads remain unpolished. The thread-ends come up flush with the polished surface of the cap jewel so these are rounded off and burnished to a high gloss. The screw that has the polished head is the escape bridge cap (upper) jewel screw as this is fastened from the *top* of the bridge, down through the cap jewel into the bridge.

The unpolished screw is the lower cap jewel screw. This screw's head is covered by the dial and its thread-ends concealed by the plates and the balance. Because this screw must support the cap jewel alone, it is often made a little heavier than the upper cap jewel screws. All this is shown in Figure 9.

Sometimes, there are two lower cap jewel screws that are similar except for length; one being very short. The shorter screw belongs to the lower escape cap jewel. If the longer screw is used here, the watch will not run. Figure 10 shows why. Here the longer screw emerges through the lower plate and comes up between the spokes of the low-lying escape wheel and acts as a block. The longer screw should be used to bind the lower balance cap jewel.

Setting bridge screws are heavier than cap jewel screws because they must fasten a heavier piece and space for a larger screw is made available.

The longest of the thin-small screws is that which sets the stud in place. This stud screw is often mistaken for a jewel screw. It is thin because there is not much room in the balance bridge's thickness for a heavier screw. What it lacks in thickness it must make up in length. This screw is of the "set-screw" type as it binds or sets against the stud. It should not be *too* long or it will

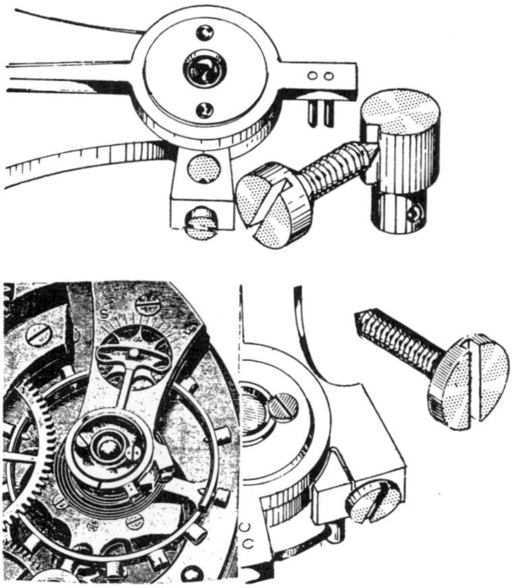

Fig. 11. Types of stud screws. Because of the necessity for having thin heads, some types, as shown in top illustration, have a slight bevel under the heads to lend strength.

reach out beyond the diameter of the movement and will be broken off when the movement is fitted to a case. Its head must be of small diameter so that the sides of the screw head do not make contact with the balance rim or possibly touch the lower coils of the hairspring should it be used in a Breguet type watch. Some stud screws have a slight bevel under their heads to lend strength and prevent the thin head from breaking off with the twisting pressure of the screwdriver. All these are shown in Figure 11.

CLICK SCREWS

The weakest screw from the point of construction is that which holds the click. It must be wide enough to cap the click. Its head must not be too high or it will interfere with the closing of the streamlined case back or rise above the rest of the movement. Because the click is posted around a hollow threaded pipe, the threads inside this short boss are necessarily thin. Most click screws look like that shown in Figure 12A. Manufacturers of watches with clicks and screws of this type do not suggest the removal of this part when the watch is to be cleaned as it may lead to weakening of the thin threads and screwhead.

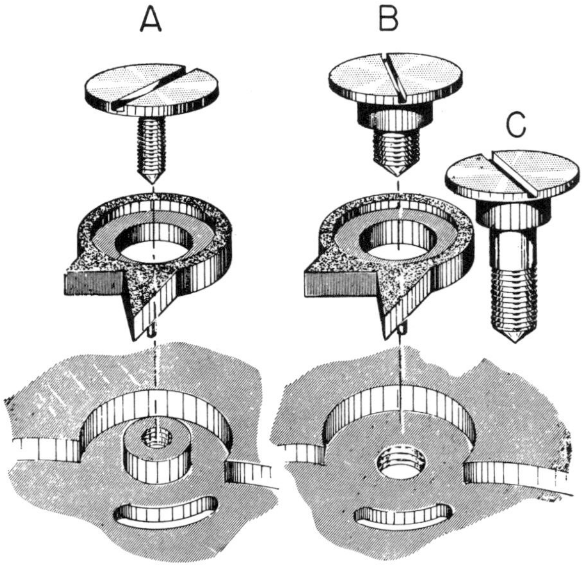

Fig. 12. Three types of click screws. Most common type as shown by A, B and C are shoulder types, the latter is larger as it is designed to serve the dual purpose of a bridge screw.

Sometimes the movement has no threaded pipe or boss upon which the click may swivel. In such cases, the screw must be designed to be able to bear all the thrust of the wound mainspring via the ratchet wheel. It also must provide a surface upon which the click will swivel easily and without binding. It must have a

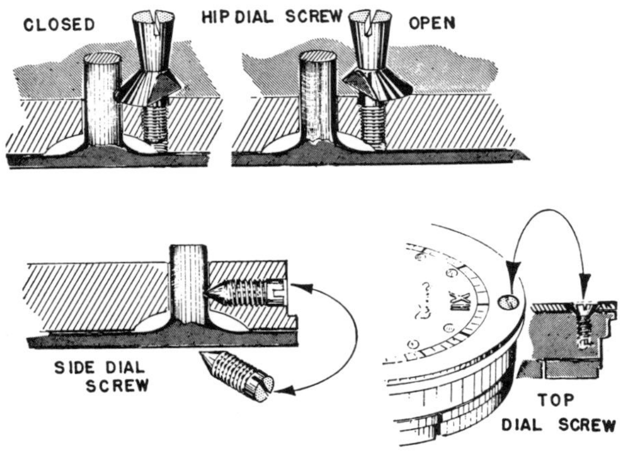

Fig. 13. Dial screws are of three types—hip, side and top.

head to cap the click and a shoulder to act as an arbor upon which the click may turn, and its threads must be thick enough to withstand all the strain put upon it by the ratchet wheel. This screw is pictured in Figure 12B and is of the shoulder-screw type. Such a screw is often used to hold a clutch lever, minute wheel, crown wheel or any other turning or swinging part.

When the clickscrew must serve as a bridge screw as well, its threaded portion does not begin until its lengthened shank reaches the lower plate as shown in Figure 12C. These screws must have shoulders of exact thickness and height to provide the click with the proper endshake and sideshake and permit freedom of movement.

CASE SCREWS

Normally, the longest screws in the watch are the case screws. These must go through the top of the movement near its edge and thread into the lower plate. Their threads are as heavy and as long as possible to withstand the strain of pulling the movement up tight. The threads must be thin enough however, to remain within the edge of the movement. The heads must also be wide enough to reach out from the edge of the movement and overlap the rim of the case. The tightening action is to pull the movement up against the case frame so that the top front shoulder is snug against the front of the case.

Occasionally, of the two case screws, one is a half-head. When a watch comes equipped with one half-head and the other a full-head screw, the half-headed screw should be placed in the hole farthest from the crown and stem. This is done so that when the movement must be removed from the case for minor repairs or adjustments, neither of the screws need be removed. In this event, the full-head screw is loosened one or two turns and the half-head screw just turned so that its straight side is streamlined with the edge of the movement.

In some water-resist watches, the movement must be pressed forward from the inside of the case toward the front. Here, the case screws are short and their half-heads fit into a recess in the inside wall of the case. To force the movement forward, these screws are turned outward, acting as lifting jacks. Screws similar to these are shown as the "D" screws in Figure 3.

DIAL SCREWS

Actually, there are three kinds of dial screws; hip, side and top.

Hip dial screws are shown in Figure 13. These have sharp hip-like edges that cut into the sides of the dial legs. These are shown both in the locked and open position. Side dial screws are another form of set screw, working from the side of the movement and whose sharp screw-ends set into the side of the dial leg as shown in Figure 13. These may or may not have a head, the screw slot merely being cut into the threaded section. Top dial screws are used to secure the dial from the top of the movement through the dial and thread into the movement as shown in Figure 13.

Hip dial screws are used where it is desirable to remove the dial

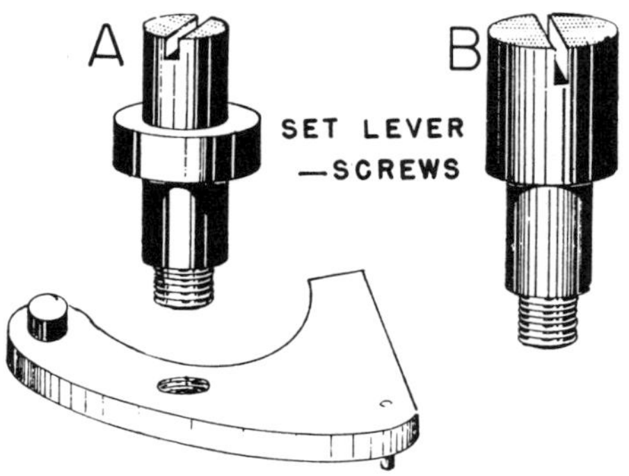

Fig. 14. Two types of set lever screws. Head of B may appear as any other bridge screw, so to indicate difference it may be colored blue.

while the movement is cased (in a 3-piece case). Their drawback is that their threaded section is necessarily short and cannot give great security. Also, their sharp edges mutilate the dial leg so that eventually, after several removings of the dial, it can no longer remain secure. Side dial screws, however, do a better job of securing the dial without mutilating the dial legs. Where a dial is too thin to support legs without danger of crimping the dial, top dial screws are sometimes employed.

SET LEVER SCREWS

Of all the screws in the ordinary watch, the set lever screw is the most complex. Its thread must fit the set lever. The lower shoulder must be long enough to emerge just a little bit through the lower plate and provide shoulder clearance so that the lever does not bind on the main plate. Its thickness must be just right to provide a minimum of side shake. Its middle shoulder acts as the screw's head as well as the controlling element of endshakes.

It must fit the recesses provided for it with very close tolerances. The upper post acts as an extension so that it reaches the back surface of the movement and is slotted to be available to the screwdriver.

There are two general types of set lever screws. These are shown in Figure 14A-B. When type "B" is used, its head may appear as any other (bridge) screw on the movement. To indicate its position and difference, this screw may have a blue colored head.

Chronographs, stop watches, calendars and repeaters, all have many different types of screws, some look like screws but have no

Fig. 15. Friction-type, off-center screw is used in chronographs, calendars and repeaters to adjust depthing between wheels or a cam and lever.

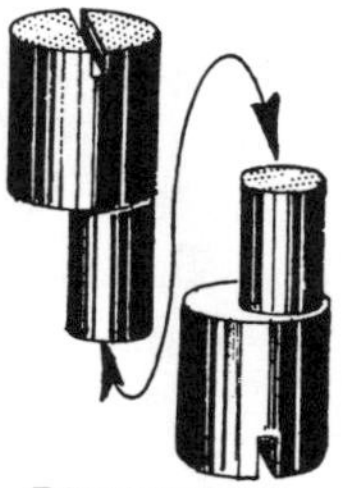

threads, merely fitting friction tight in the plates. Their heads are off center (eccentric) and are used to adjust the depthing between two fine wheels or the depth of engagement between one cam or lever with another. Such a screw is shown in Figure 15. For the greater part, most of these special types are much the same as ordinary screws or a combination of them.

BOOK II

CHAPTER II

Part II

HOW TO MAKE A SCREW

ALTHOUGH almost all watch parts are available today, occasionally it is necessary to make some parts of the watch. This is especially true of screws. Some watchmakers who possess bottles of assorted screws often find themselves in a position where they must make a different kind.

The knowledge required to make a screw encompasses more than the use of the screwplate. The diameter and height of the screwhead, the type of fit, the bearing surface if any, the direction and character of the thread, the polish and temper of the screw, all enter into the watchmaker's finished product.

For a sample project, the making of a set lever screw will be undertaken and described. The set lever screw (Figure 1) is selected because this screw contains practically all the parts of every type of screw used by the watchmaker.

(1) It has threads which turn into the hole of the set lever (H-J, Figure 1); (2) it also has a post which fits through a close fitting hole in the lower plate and extends below it just enough to provide clearance for the set lever (G); (3) it has a retaining shoulder (A-F), most often in the middle of the screw's length which governs its endshake. The post above the shoulder usually extends upward until it is level with the upper plate. This shoulder (B-E) is slotted at the top so that it receives the blade of a screwdriver handily.

Let us suppose that the original screw has been lost. In making a replacement, we must use the parts which it fits as the guides for its construction.

To make the screw, a piece of annealed high carbon steel or drill rod is used. (Phonograph needles, dentists' burrs, or the shanks of old drills, annealed, make acceptable material.) The thickness of the metal for this screw must be at least as thick as the hip (A, Figure 1), which is the thickest part of the screw.

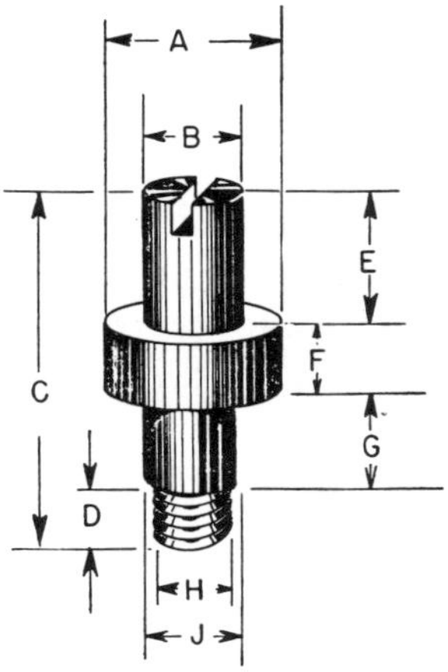

Fig. 1—Some of the measurements to be considered in making a set lever screw.

Since we must proceed without a sample, we get this gauge from the countersink which is on the inside of the barrel bridge (shown in Figure 2). This recess is used to accommodate the retaining shoulder (A, Figure 1). The metal is cut cylindrically until the flat blunt end of the stock fits snugly into this recess and long enough to provide for the entire length of the screw.

The next operation is to reduce the thickness of part of this cylindrical section until it fits through the hole in the *lower plate.*

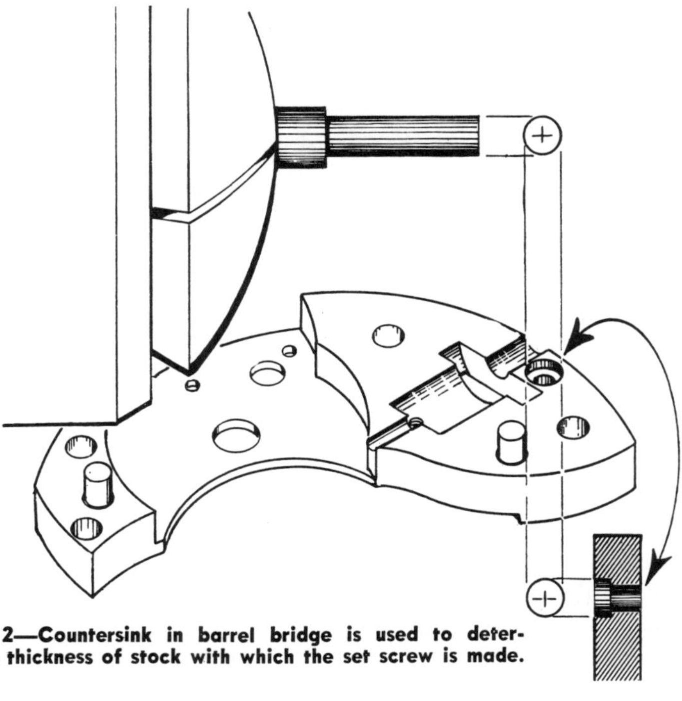

Fig. 2—Countersink in barrel bridge is used to determine thickness of stock with which the set screw is made.

This new post must fit snugly, otherwise the screw might have too much sideshake. It also should be cut long enough to extend through this hole a short distance but sufficient to have threads cut upon it to accommodate the set lever.

A shallow groove is cut upon this new shoulder at a point just above the area where the metal extends from the lower plate (shown by arrow in Figure 3). The space shown by the two smaller

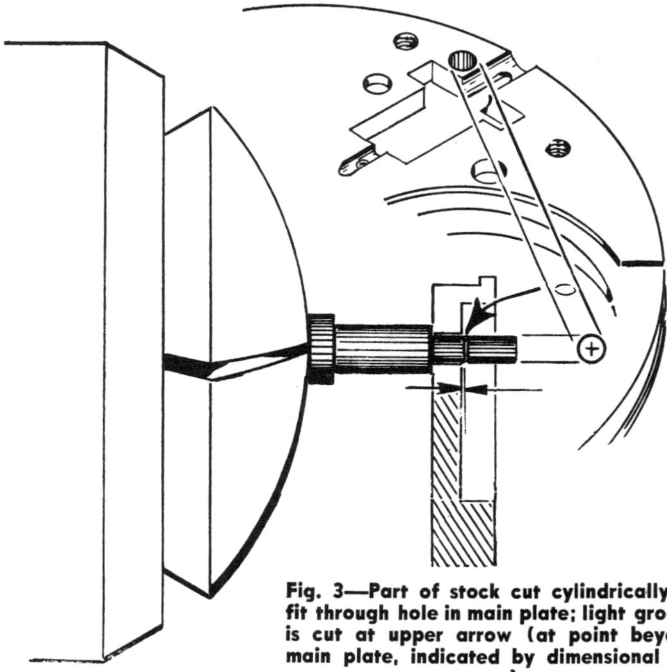

Fig. 3—Part of stock cut cylindrically to fit through hole in main plate; light groove is cut at upper arrow (at point beyond main plate, indicated by dimensional arrows).

arrows indicates the clearance above the plate where the groove should be cut. This groove marks the lowest position on this extension where the threads may be cut. If the threads are cut lower than this groove or position, the set lever may be pulled down too far and bind upon the lower plate. Of course, if the threads are not cut low enough, the lever may not come sufficiently close to the movement to grasp the stem slot. In such a case, the stem will not be secure in the movement.

The thickness of the threads is determined from the screw hole in the set lever. A piece of soft brass is turned until it can be *pushed* through this hole. This size is the base diameter of the screw. Add one-quarter of its size to obtain the full thickness of the threaded section. Let us say, for instance, that the unthreaded wire fitting through the set lever hole gauges .68 mm.; 25% of 68 is 17. This added to .68, totals .85 mm., the diameter of the screw to fit the set lever (screw) hole.

SCREWPLATES

Threading this section of the screw is done with a screwplate (some models of which are shown in Figure 4). There are many types of screwplates available. Some are designated by letters such as B, E, F, G, L, M, S, etc. These letters usually suggest a difference in the distance from the crest of one thread to the crest of another ("pitch"). Some have fine threads used for watch screws, some are heavier for use as optical screws, and still others are coarser for jewelry. The numbers shown on the side of these

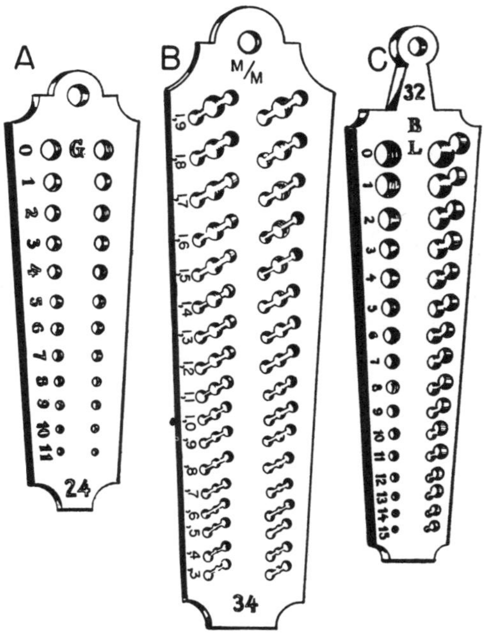

Fig. 4 — Screwplate "A" is plain; "C" single-matched with a row of plain holes; "B" is double-matched with metric graduation.

screwplates run invertedly from "O," their largest diameter screw-hole, downward to 1-2 and continuing until 20 or 21, becoming progressively smaller. (The higher the number, the thinner will be the screw it will cut.) Screwplates of German or English origin bearing the letter "L" on the screwplate, or this letter added to another, denotes that it is a left thread. Some French or Swiss screwplates that are left threaded are marked with the letter "G" for "gauche" (left). Today most screwplates are in metric gauges.

Sometimes, two screwplates made by different companies and marked with the same letter vary as to the number of threads each will cut to a millimeter. Furthermore, the diameter of the numbers, 0, 1, 2, etc., do not always conform. On one screwplate, the number "0" may cut a thread gauging 2.20 mm., and on another the thread cut on the "0" hole may gauge 1.80 mm. If one need be purchased, it is recommended that a screwplate with millimeter markings such as shown in Figure 4-B be obtained. The hole sizes marked indicate the outside thickness of the threads it will cut in tenths of a millimeter. Thus, a hole marked 9 will thread a screw whose thickness will be .90 mm.

Screwplates come in three styles: plain (A, Figure 4), combination plain and single cut or "notched" (C, Figure 4), and double notched (B, Figure 4). The best of these three is the double notched.

When the plain "cut" is used, the threads are squeezed rather than cut into its surface. This requires more force to thread and often results in breaking. Also, when such a screwplate is used, the screw blank must be prepared a little *thinner* than the eventual thickness planned for the threads. With this screwplate, should an .85 mm. thread be desired, the prepared blank should gauge .80 mm. This is suggested because in forcing the blank through the thread die, the sharp edges displace rather than cut the metal. The displaced metal is literally squeezed into the crests of the threads in the die—and the result is a screw with a diameter greater after threading than before threading. This is also true of worn or old screwplates.

With the single cut (C, Figure 4), and better with the double cut or notched (B, Figure 4), the threads are cut rather than knurled. The cutting edges, at either side openings of the die, cut the grooves which form the bottom of the screwthreads. The number (34) at the bottom or top of the screwplate gives the number of the holes in the tool. Most often, these are supplied with a double row of holes. One row contains the full or even sizes, such as 9 (.90 mm.), 8 (.80 mm.), etc., while the opposite row has the half sizes, .85, .95, 1.05, etc.

A method of determining the thread sizes has been given previously; however, with the mm. screwplate, this is simplified. The pin that originally was turned to fit (unthreaded) into the set lever screwhole can be fitted through the hole in the screwplate in the same way. The full diameter will be found marked in tenths of a millimeter next to the hole into which the pin fits closest. Should the pin fit snugly through the number 8, the diameter of the screw will be .80 mm. The screwblank should gauge this size.

It is a good practice to make the threaded section a little longer than needed. This provides for some imperfection in the threads which occurs until the screwplate has sufficient metal in it to cut evenly. Also, to permit easy entry into the screwplate, a slight taper should be cut on the beginning of the portion to be threaded. When the threading is done, the taper and excess length may be cut away, leaving the screw just long enough to emerge perceptibly through the set lever.

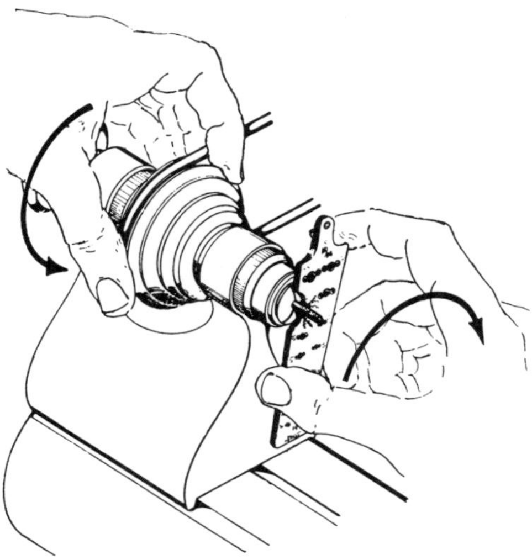

Fig. 5—Threading is done in gradual cuts with lathe head turned only by light hand-twists.

Actual cutting of the threads should be done while the stock is held in the lathe. The screwplate is held on its sides between the index, middle finger and thumb of the right hand while the left hand grasps the headstock lightly and supplies the force needed to rotate the stock through the screwplate. This manipulation is shown in Figure 5.

A little lubrication is desired during the threading operation. The threading should be done by turning the stock with the screwplate about ¼ turn and then reversing this action at least a full half-turn. This will cause the shaving to break off, leaving the next ¼ cut easier to take. It also will prevent these shavings from clogging the work and spoiling the threads.

The stock and screwplate are twisted towards each other again until the "feel" indicates that further force may either strip the thread or break off the screw. Usually, advancements of ¼ turn should be safe and insure against failure. This is continued cautiously until the threading is done.

The screwplate is removed from the screw carefully, lest this movement strip or tear the threads. If the work is cleaned and cleared of shavings occasionally as the threading progresses, the danger of "tearing" a thread will be practically eliminated. Cleaning the threads after the screwplate is removed is done by pressing a piece of cork against the threads while the lathe motion is reversed.

The threads should be cut up to the groove previously made to indicate the lowest thread for the set lever. The lever is then threaded on the screw, testing for fit and closeness to the groove.

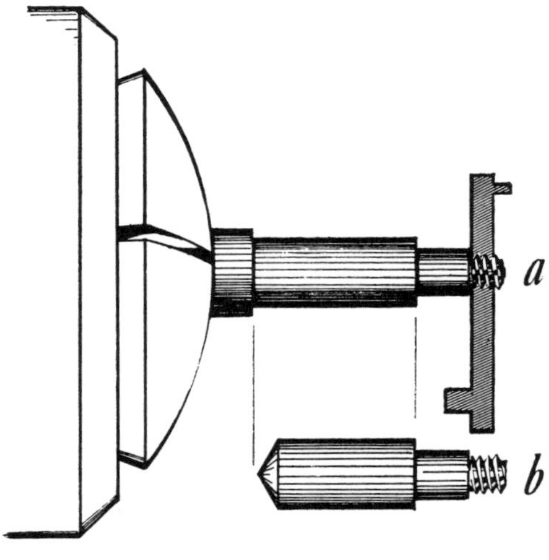

Fig. 6 — Set lever is tested on the nearly cut thread (a). After surplus thread is removed the blank is severed as shown (b).

All excess threads but one, extending out of the lever, should be removed (Figure 6, "a"). This leaves a safe margin and permits the thread to be short enough so that eventually it does not interfere in any way with the dial.

Next, the post fitting through the barrel bridge is cut. To do this, the blank is cut off at a height just above where it will reach through the barrel bridge (Figure 6, "b"). The extra length is required to compensate for the later grinding and polishing of the head. This is done because the blank, as far as it is finished, will have to be held in the lathe chuck on the lower shoulder (Figure 7). If too much stock is left out of the chuck, strain on the stock held in the chuck may result in difficulties in cutting and inaccuracies in fitting.

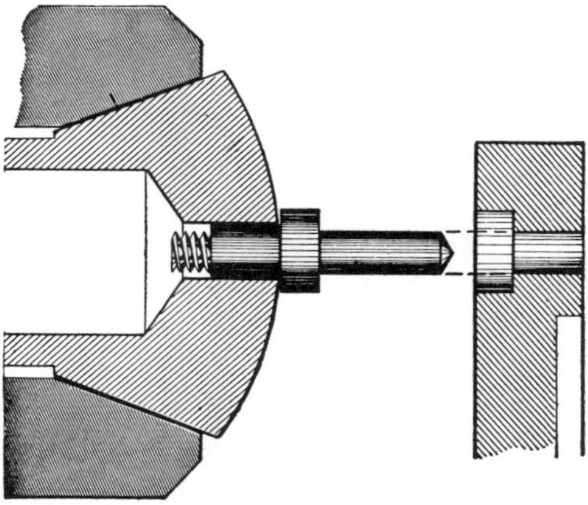

Fig. 7—After the blank is reversed in the lathe, a cylindrical portion is turned to fit through the hole in the barrel bridge.

After the blank is reversed in the lathe, held by a well fitting chuck, a portion of the end is reduced in diameter until it fits snugly into the hole in the barrel bridge. (This establishes the correct diameter of the shaft of the screwhead). The shaft may be made longer by cutting down the shoulder of the collar until the shaft fits through the hole in the barrel bridge and the back shoulder of the screw is flush with the edge of the countersink in the barrel bridge.

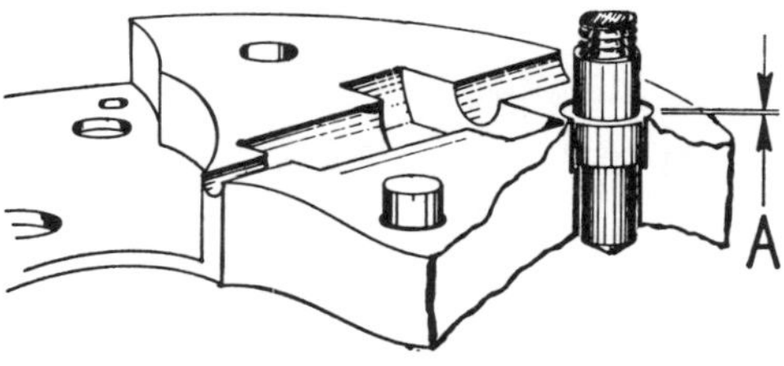

Fig. 8—Hub is cut until it fits below counterbore (amount shown by lines "A" provide for endshake).

As shown by the arrows in Figure 8, the shoulder may be slightly below the level of the bridge (the amount indicated), allowing for the maximum of endshake permitted. All shoulders must be cut square and no *rounds* may exist on the inside corners (round corners give a false fitting which will in a short time work loose, and then the stem itself will easily fall out of the movement).

Excess length which extends through the barrel bridge may be trimmed off so that the end is almost flush with the plate, leaving just enough for grinding and polishing. (This should be checked while the stock is still in the lathe). The barrel bridge is fitted over the stock, and the amount of metal protruding is noted for removal.

The next operation is slotting the screwhead with a screwhead file (Figure 9). If such a file is unavailable, a suitable slotting saw may be made from a single-edged razor by placing the razor lengthwise over a No. 4 file and hitting its back with a sharp blow of a hammer. The sharp edge of the razor will assume the profile of the file's surface, and these indented teeth will slot a screw of moderate temper.

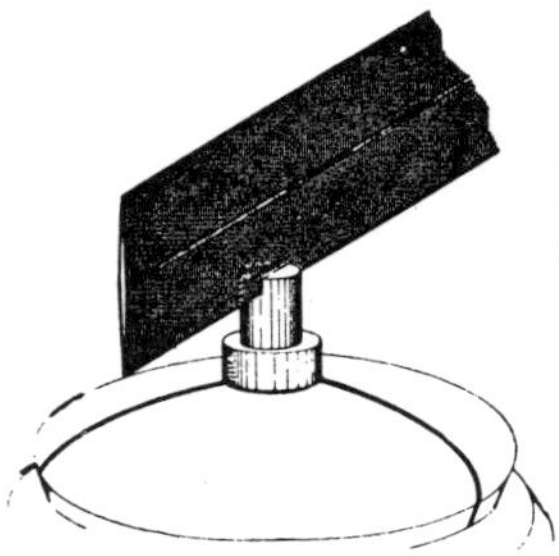

Fig. 9—The screw slot is cut with a screwhead file, while the blank is held in the lathe or pinvise.

The screw now may be tried in the movement and tested for endshake, which must be barely perceptible and with practically no sideshake. However, a set lever screw that is too tight between the plates will cause difficulty in pulling the stem into the setting position and may loosen the set lever when it is pushed back into the winding position. Conversely, one that is too loose may cause the stem to fall out.

POLISHING SCREWHEADS

When the screw is slotted deeply enough to permit sufficient depth to remain after grinding and polishing, the screw should be hardened and tempered to a blue color and the head polished. If the head is to be slightly rounded, the polishing is done with the screw in the lathe. An oilstone slip may be used to round off the head, followed by the application of a jasper stone to remove scratches caused by the grinding action of the oilstone. A final application by a burnisher will give the screwhead a high gloss finish.

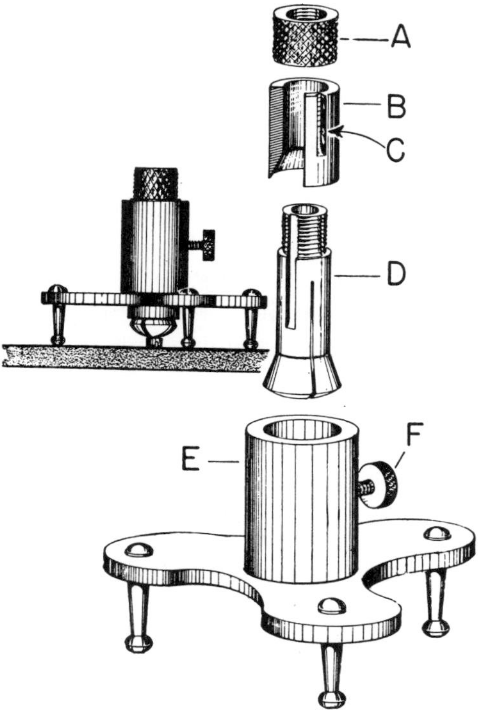

Fig. 10—Grinding and polishing may be done effectively with this tool (assembled unit is inset—upper left, opposite "D").

If the screwhead is to be polished flat, then a different and more detailed process must be followed. The screwhead must be ground flat in preparation for polishing. Both the grinding and polishing operations may be done with screwhead polishing tools or fixtures. There are many different types of these devices. One of the most efficient is illustrated in Figure 10.

This tool consists of a regular lathe chuck holder set upright into a 3-legged platform. The knurled nut "A" draws the chuck "D" up through the holder "B". This little unit is then fitted into the pipe "E." The chuck and holder are permitted free *vertical* movement, but the nut "F," working in the groove "C" of the chuck holder, prohibits any twisting motion. The unit with the chuck holding the screw is placed over the grinding surface. The screw with the weight of the holder over it bears on this surface. The assembled tool is shown in the inset opposite the chuck "D" in Figure 10.

To grind the screwhead flat, a mixture of oilstone powder and clock oil is made into a paste the consistency of cream cheese. As much of this paste as can be scraped with the outside of the thumbnail is transferred to a piece of flat frosted plate glass about 6 inches square. The screwhead tool with the screw is placed over the mixture on the glass, and the tool is rotated in circles or ellipses of about 1½ or 2 inches. The index finger presses upon the nut holding the chuck, while the thumb and middle finger straddle the holder tube during this operation. The action of the tool over the abrasive-covered flat glass will flatten the end of the screwhead and ready it for the final operation of polishing.

The actual polishing is similar to the grinding operation except that polishing powders are used; however, the screwhead and the tool must be cleaned and dried perfectly so that not a grain of grinding powder remains to mingle with the polishing compound and spoil the polishing operation. The screw should not be removed from the chuck or tool while it is being cleaned, or the position of flatness in its relation to the legs of the polishing tool will be upset.

The polishing compound may be fine diamantine mixed with watch oil into a paste of the same consistency as the grinding element. As in the grinding operation, a very clean piece of frosted plate glass is used and the tool manipulated in the same manner. Cleanliness in polishing is a *must* if a successful job is to be expected; no foreign matter should be permitted to be introduced.

For the very small area of a screwhead, little time should be required to grind it flat and bring the head to a black flat polish.

Jewelers' rouge may be used as an alternative to diamantine in polishing the screwhead. This will give a quick and brilliant finish if the rouge is mixed with a little alcohol (instead of watch oil). The alcohol evaporates quickly but the drying rouge "grabs" the steel surface and brings out a good lustre readily. A few drops of alcohol are dropped on the plate glass, and a clean bar of rouge is rubbed on the wet surface, causing a little rouge to adhere to the glass. The polishing tool holding the screw is applied to the glass, and manipulated in the same way as in grinding. While the screwhead is being polished on the rouged surface, the alcohol will dry, and the tool as it is moved over the drying surface may cause a squeaking sound. (This should be considered as part of the process of a successful job). The screwhead is then inspected, removed, and cleaned.

Grinding and polishing removes some of the metal. This should be considered when the screw slot is cut—so that it does not become too shallow by successive grinding and polishing.

TAPPING HOLES

While the preceding paragraphs explained the making and finishing of a screw, occasionally a genuine replacement screw will not fit the threads in its screwhole because a previous unsuited screw may have been forced into the hole, stripping the threads. The hole will have to be re-tapped with a slightly wider thread and fitted with a specially made screw having a thread diameter and pitch the same as that of the tap. The watchmaker will often encounter this situation in the threaded holes for the crown wheel in the barrel bridge.

The first step in tapping a hole and fitting a screw is to determine the larger diameter of the screw required, following the procedure previously explained. Using a shaft considerably longer than the length of the finished screw, thread the long shaft.

This longer thread is used both for tapping, and later a part of it for the screw (the tapping section is cut off after the hole is tapped).

For example, to make a crown wheel screw of which the threaded section of the finished screw must be 2.50 mm. long: To prepare

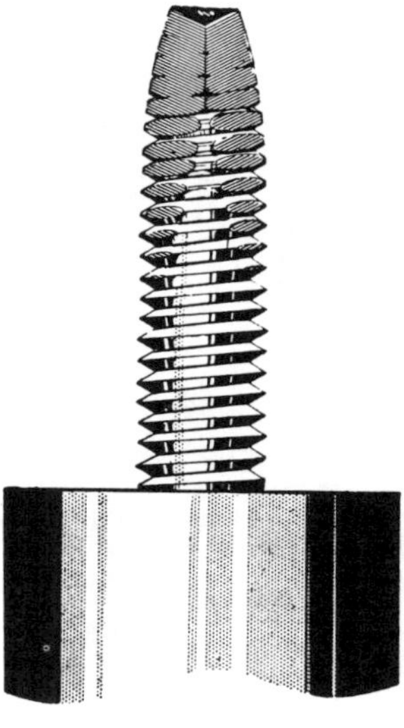

Fig. 11—A screw tap is made by cutting a longer thread, and then filing 3 sides to a taper.

the thread for its dual purpose, make it 7 mm. long. Then on the front 4 mm. section, file 3 equally divided tapered flats so that the screw appears like that shown in Figure 11.

This is then tempered and used to tap the hole, making sure to ease the tapered front into the hole so that the threads are cut cleanly. The "tap" should be threaded into the hole in the same cautious manner that a screw is threaded into a screwplate. The tap should thread completely into the hole so that the lowest, untapered (full) threads enter the hole. After this the taper part of the "tap" is removed, the remaining threads being part of the screw. In this way the newly threaded hole will have a screw that is perfectly matched.

LEFT THREADS

Sometimes it is necessary to cut a screw or tap a hole with a *left thread,* often used on crown or ratchet screws. Screwplates with threads tightening in the counterclockwise or left direction are available at material supply houses. When purchasing these, make certain that the holes are graduated in metric thread diameters, rather than by some strange system.

It is possible to make a left thread by using a right (clockwise) threaded tap or die. To do this, a steel thread long enough for tapping purposes is made. Then with a fine file remove the threads on 178° of each side of the threaded section. This will leave

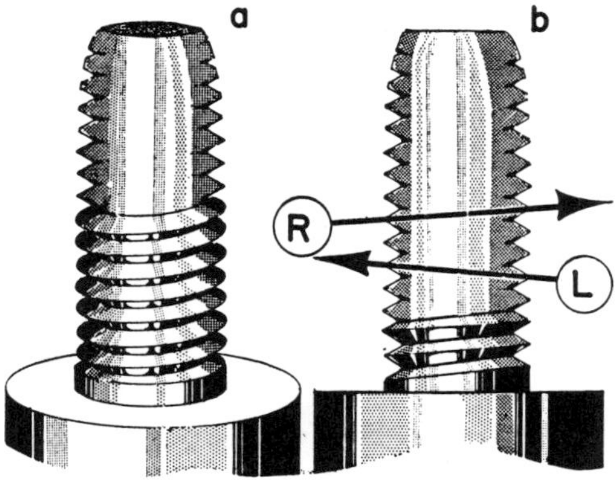

Fig. 12—To convert a right thread to a left, a long screw is filed to canoe-shape (elliptical). Finished tap is shown in "b." This will cut both right and left threads.

2° of threads diametrically opposite each other. (If the cutting edges are tapered slightly as shown in the accompanying drawings, the tap will cut threads in either direction.) The tap should then look like the one illustrated in Figure 12, designated "a"; and in profile, it will resemble the one in Figure 12, marked "b".

It can be seen in profile that the crest of one thread is (as in any direction thread) opposite the base thread on the other side of the tap. This can either be a descending or ascending base. This is true whether it be the profile of either a left or right thread. Therefore, if this profiled tap is pressed slightly as it is introduced into a hole, it will thread in whatever direction it is turned. Of course the tapping should be done cautiously and in easy stages, frequently clearing the hole to remove scrapings and chips.

Fig. 13 — To make a left-threaded screwplate, the tap in figure 12 is used to thread this plate; after 2 bordering holes are drilled, the metal between the holes is cut through with a jeweler's saw.

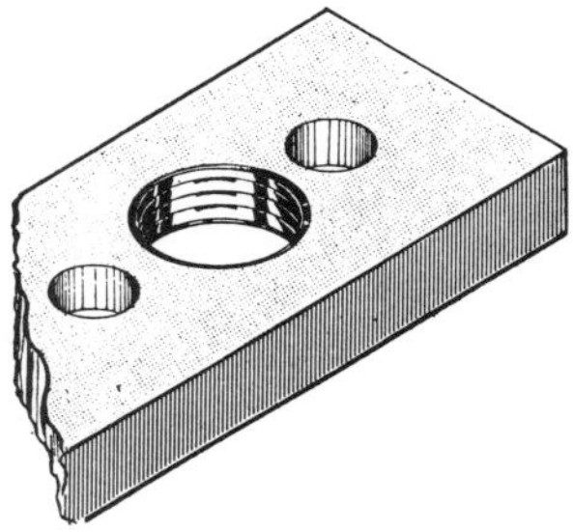

To make a die for left threading screws, the same type of (tempered) tap is used, but threading should be into a plate of annealed high carbon steel. Two holes adjacent to this threaded hole are drilled (as shown in Figure 13). Using a jeweler's saw, guide the blade through the outside hole and cut a slot so that all 3 holes are joined to make it appear as a double notched screw-plate hole (as shown in Figure 4B). Subsequently the plate is hardened and drawn to a light straw color and then is ready for use.

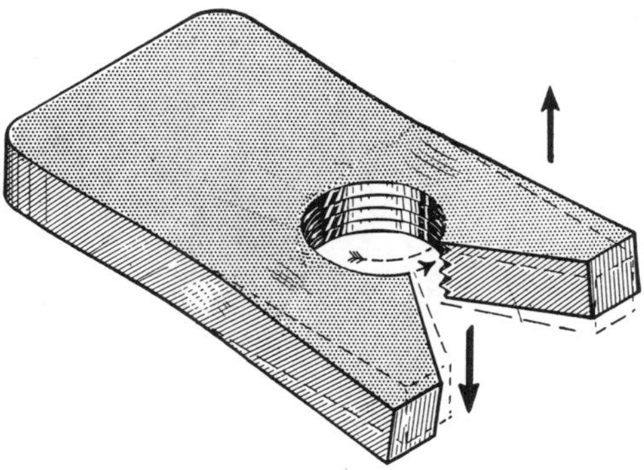

Fig. 14—Left-thread screwplate can be made by slitting right-thread hole and twisting prongs as shown to change thread direction.

Another system used to convert right threads to left threads was described in the Swiss *Horological Journal.* A threaded hole in a piece of annealed stock is cut open at one end to provide for cutting rake. Then the left prong of the hole is bent downward and the right prong bent upward so that the twisting direction of the threads is changed (as shown in Figure 14.) After this is hardened and tempered, it will cut fairly clean left threads.

CLEARING OUT CLOGGED SCREWPLATES

Holes in screwplates may become clogged with broken-off screws, rendering the screwplate inoperative for these holes or sizes. A simple method of removing these obstructions is to take a jeweler's saw, disconnect one end of the blade from the frame and after

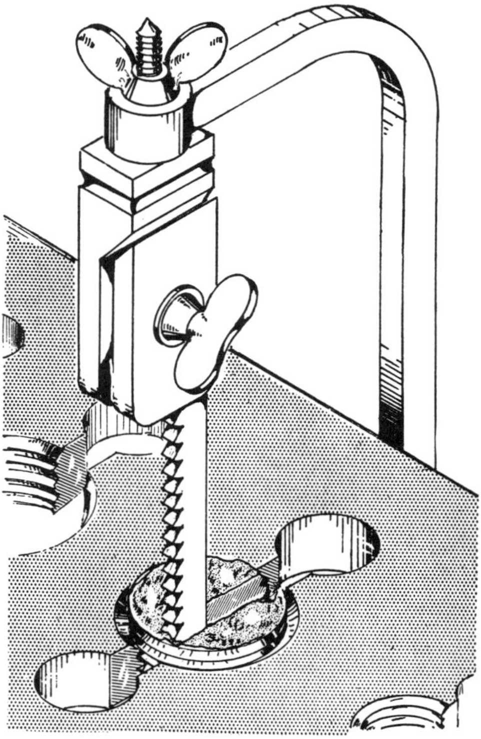

Fig. 15—Clogged screwplate can be cleared by cutting through the screw with a jeweler's saw.

threading it through one of the guide holes, fasten the loose end again to the saw-frame. Thus the saw is in position for cutting across the clogged screws (as shown in Figure 15), when it can be easily removed. The saw blade will not harm the threads in the screwplate because these saws are not tempered quite as hard as the metal of the screwplate.

BOOK II

CHAPTER III

USES OF STAKING SET STUMPS

A little used part of one of the watchmaker's most versatile tools, the staking set, are the stumps supplied with each set. Of various sizes and shapes, these stumps are each designed for a specific use.

MANY watchmakers often have jobs calling for an increase of the endshake in a barrel, the staking of rollers or staffs for old 3-arm balances, hole closing, and other odd jobs. To perform these tasks they resort to the use of makeshift tools or methods, not realizing that in their own staking sets are stumps that were made specifically for the job they intend to perform.

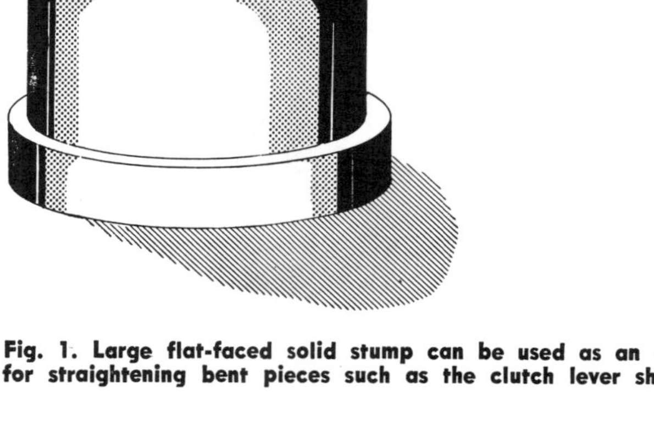

Fig. 1. Large flat-faced solid stump can be used as an anvil for straightening bent pieces such as the clutch lever shown.

Staking sets have always been supplied with many stumps of various sizes and shapes, each one to serve a specific purpose. However, they are included in these sets with the premise that the buyer has a full knowledge of their use. Possibly for this reason no manual of instructions is included. It is intended here to describe the various stumps and their use.

Generally, in the modern staking set, the punches go through different tempering processes. Not all punches are of uniform hardness. For instance, a cross-hole punch, used in conjunction with a roller remover and in other staff work, is annealed after hardening to a much greater extent than a round-faced solid punch. Mouth tapered punches also are not as hard as riveting punches or the flat-faced punches. These facts must then be considered.

Because stumps must often act as miniature anvils and take a

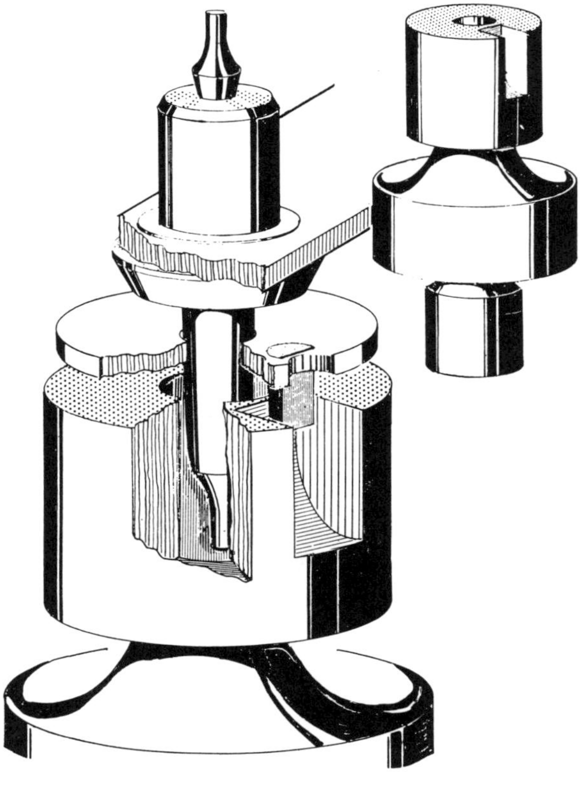

Fig. 2. This stump is used to stake on single rollers and lower part of two-piece rollers.

great deal of abuse, they are left much harder after the necessary tempering process is completed.

To use the punches unnecessarily as stumps when a stump is handy, invites the shattering of the punch because it was too brittle, or spoiling it because it was of necessity, made softer for the sake of toughness.

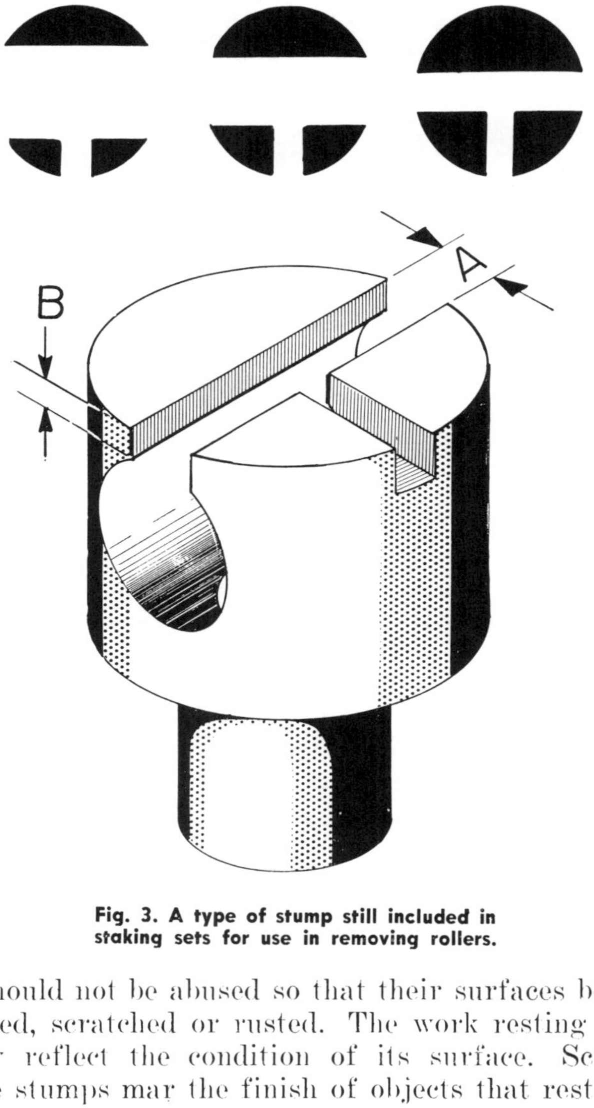

Fig. 3. A type of stump still included in staking sets for use in removing rollers.

Stumps should not be abused so that their surfaces become badly nicked, dented, scratched or rusted. The work resting on a stump will directly reflect the condition of its surface. Scratches and dents on the stumps mar the finish of objects that rest upon them. Rusted or damaged flat stumps may be refinished with a screw-head polishing tool.

When using flat-faced stumps, it is advisable to use one with the largest diameter the work will permit. In this way, the work rests on a large, flat surface, giving it stability and easier manipulation.

In Fig. 1, is shown a high grade clutch lever resting upon a flat-faced solid stump in preparation for taking out a bend. With a stump of this diameter, it is possible to rest the piece on the flat, raised table without the possibility of it falling off or shifting while the lever is being pressed. Where possible in the use of flat-table stumps, the largest size available should be utilized.

In Fig. 2 is shown a stump used to stake on single rollers and the lower part of two-piece rollers. Staking sets come equipped with these stumps graduated in hole size, outside diameter, and also in width and depth of the slot which accommodates the jewel.

In selecting a stump for this operation, the staff's roller post must fit into the center hole of the stump with ample room around it. The stump must also have its slot wide enough and sufficiently deep to accommodate the roller jewel. The slot must also be cut close to the center hole in the stump, so that the roller jewel will have clearance from the rear as well as from its sides and depth. Otherwise this jewel might crack or break when the roller is pressed downward.

To use this stump, the roller table or disc is placed upon the staff with the tweezers, pushing it down far enough so that it will not fall off when the balance is turned over. The stump is placed in the staking die, centered and then locked in place. The staff and roller are then placed over and on the stump. A flat-faced hollow punch is then fitted over the hairspring shoulder of the staff and as the hammer taps the punch lightly, the balance is pushed down onto the roller so that when the job is done, the roller rests snug against the balance staff hub.

Some watchmakers prefer not to use the roller stump for this operation, but rather a flat-faced hollow stump to rest the balance on it so that the hairspring shoulder of the staff is in the hole of this hollow stump. Or by inverting the staff, and with a punch, bearing the same characteristics as the stump in Figure 2, the roller is pushed downwards.

In Figure 3, is shown a type of stump that is still included in staking sets and is used to remove rollers. While roller removers are sold as separate items and tools, staking set manufacturers still include them in order to give versatility to their product.

These roller-removing stumps are not adjustable for rollers of many sizes. That is, they cannot be made wider or narrower. To remove the roller, the balance must be manipulated so that it is edged between the jaws of the stump, and the roller rests on the stump as shown in Figure 4. Here, the roller's edges rest upon the stump, while the balance arms or spokes are below the lips of the stump. To fit the balance to the stump, it is tilted so that the balance arm, which is often wider than slot "A" (Fig. 3), may be edged in under the slot and thus permit as much of the roller to rest on the stump as possible. If the roller is supported merely at its extreme edges, it may crack or bend, and furthermore, it may be difficult to remove. This is due to the slight bending of the roller disc, causing the warped hole to pinch the sides of the roller post.

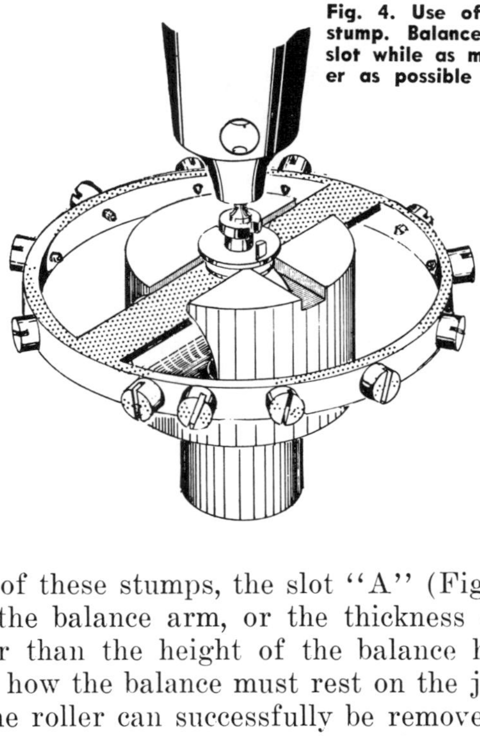

Fig. 4. Use of roller-removing stump. Balance arm is below slot while as much of the roller as possible rests on stump.

To make use of these stumps, the slot "A" (Fig. 3), must either be wider than the balance arm, or the thickness of the lips "B" must be thinner than the height of the balance hub of the staff. Figure 4 shows how the balance must rest on the jaws of the roller stump before the roller can successfully be removed. In the upper section of Figure 3, is shown a silhouette of a set of three roller-removing stumps. Notice that they vary not in diameter but rather in the distance between the jaws. This variation provides for the differences in both roller diameters and balance arm widths. The cross slot is to provide for clearance of the roller jewel, should it project slightly below the roller disc.

Probably the least used of all stumps in the staking set are those pictured in Figure 5. These tri-slotted hollow flat-faced stumps often puzzle users, especially the younger watchmakers. They are employed when rollers must be removed from old, three-armed balances. Sometimes, these stumps have six radial slots in an attempt by the makers to give them a wider range of employment. This makes it possible to use them with both two or three arm balances and sometimes with six-spoked chronograph wheels with special

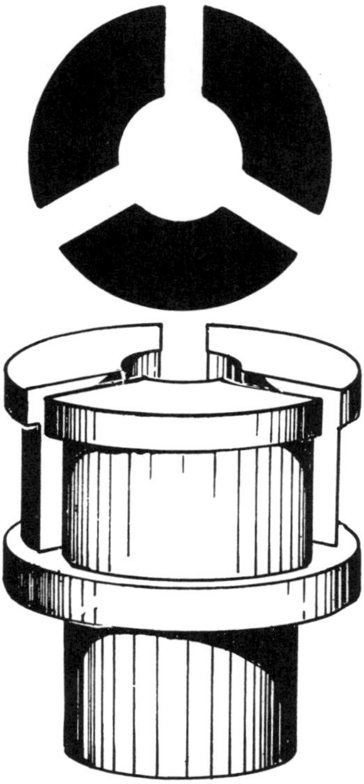

Fig. 5. Least known of all the staking set stumps are these tri-slotted ones used for removing rollers from old three-armed balances.

discs on their arbors or pinions. Figure 6 shows an old three-armed balance with a roller resting on such a stump in preparation for the removal of its roller.

The round-faced solid stump shown in Figure 7 is used to close holes in brass and steel plates or objects. These may be used as a base or stump in conjunction with a punch of similar curvature and radius when the hole to be closed does not act as a bearing, or where endshake is of no concern. Such a case may be the brass cam of a clock or a thick clock wheel that must be driven friction-tight onto an arbor or post. It is also used when a pallet fork must be stretched.

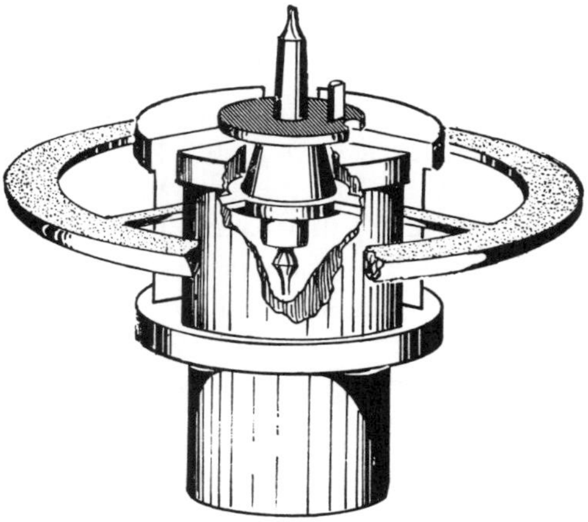

Fig. 6. Three-armed balance resting upon tri-slotted stump in preparation for removing roller.

Here the fork is rested on the central portion of the stump and a peening punch is brought over the fork and is lightly tapped. This method has less tendency to bend the fork while it is being stretched, as might be the case when it is rested on a flat-faced solid stump.

In hole-closing operations, generally, the purpose of the stump is to serve as a base or anvil upon which the work is rested while

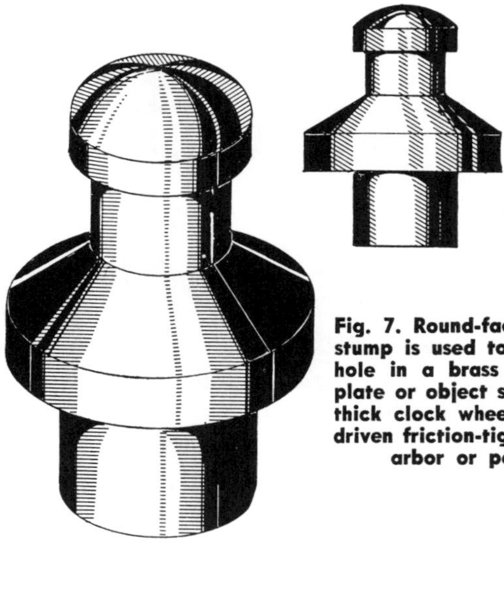

Fig. 7. Round-faced solid stump is used to close a hole in a brass or steel plate or object such as a thick clock wheel that is driven friction-tight on an arbor or post.

the punch does the job. The use of the proper stump, however, will go a long way to insure a successful job. To close a hole in a bushing or plate, the plate is rested on a flat-faced solid stump whose diameter most nearly fits into the countersink in the bottom of the plate. When such a stump is unavailable, a wide diameter flat-faced solid punch (inverted) may be used. If the bushing has an oil sink, a round punch, slightly smaller in radius than the oil-sink is used. This is recommended because its smaller size permits its surface to reach down against the hole and yet contact as much of the area in the "sink" as possible. When the punch is tapped with the hammer, the hole will close more through the depth of the

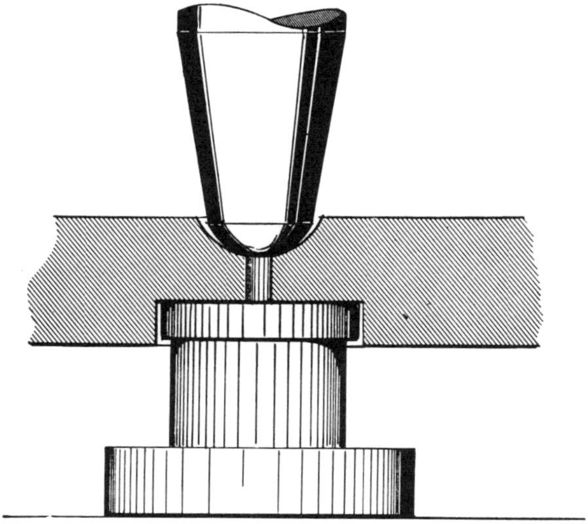

Fig. 8. A round punch used in conjunction with a flat-faced solid stump can be used to close a hole in an oil-sink as shown.

hole rather than merely pinch in the topside alone. This operation is shown in Figure 8.

Figure 9 shows a small concave stump. This tapers up to a small table, the top of which is concave. This stump is used to support the back pivot of the center wheel pinion when the cannon pinion or the hands are being replaced; or to support the back pivot of center seconds pinions while its hands are being secured. This stump is highly recommended for this purpose. To push on cannon pinions, minute or center seconds hands without the lower pivot being supported in such a stump invites the risk of damage to the jewels and bridges. Fig. 10 shows this stump in cross section while a cannon

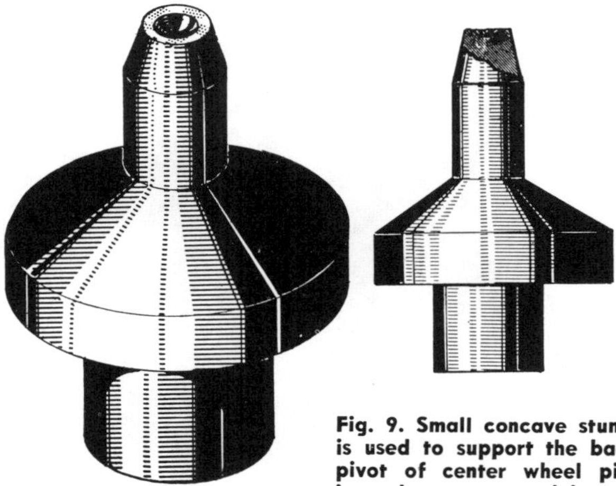

Fig. 9. Small concave stump is used to support the back pivot of center wheel pinion when cannon pinion or the hands are replaced.

pinion with a tight "knuckle" is being pushed down in place. (Also see Book III, Chapter V, Fig. 19.)

Figure 11 shows a "V" stump, sometimes called a "crotch" stump. This is used to support and center the cannon pinion while the tension dent is being renewed. Sometimes this stump is used to support a section of wire in preparation for a cross-hole being drilled through its diameter. Some staking sets have a punch with a pinvise-end attachment. These can be used to hold small pivot

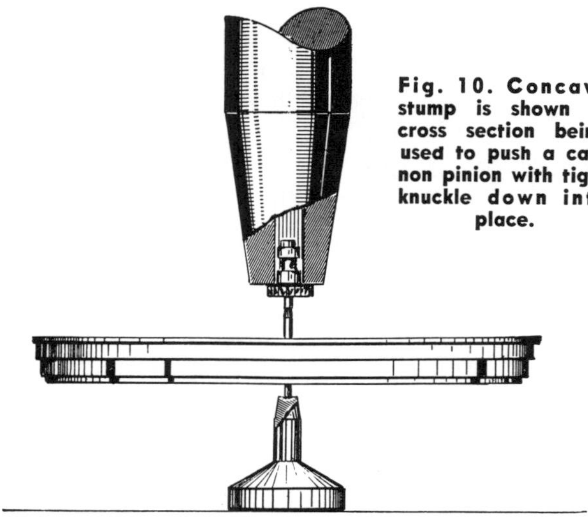

Fig. 10. Concave stump is shown in cross section being used to push a cannon pinion with tight knuckle down into place.

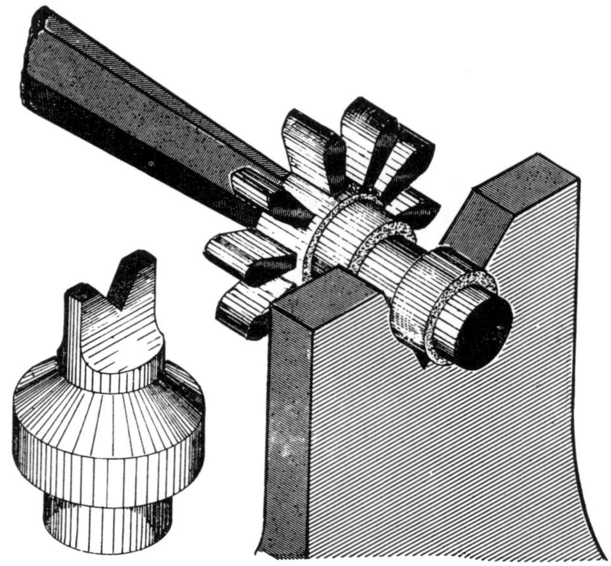

Fig. 11. V stump is shown at left and at right. Enlarged view shows how stump is used to support a cannon pinion.

drills with which the cross-hole in thin wire may be drilled—such as special stems, chronograph buttons or special small tools. (See Book III, Chapter V, Fig. 12.)

Figure 12 shows a large diameter flat-faced cup stump. This is used to increase or decrease the end-shake of barrel arbors or to "bump" plates or bridges in order to adjust the endshakes.

Much of the power generated in the barrel is lost because of the negligence of many watchmakers in failing to check the endshake in the barrel. Factory trained watchmakers consider it a "must" to place the barrel arbor in the barrel without the mainspring, replace the cover, test the endshake of the arbor, and then grasping the arbor at the square with an end-cutting pliers, spin the barrel to observe the freedom of the arbor in the barrel.

Too many watchmakers consider these operations a waste of time and then later are puzzled when the balance has a poor motion despite the newness and cleanliness of all the parts. When the end-shake must be reduced or increased, this adjustment is best made to the barrel cover. To increase the endshake, the cover is placed topside down on the cup-stump. While a flat, solid punch may do the job well, a mouth-tapered punch, fitting over the outside taper of the base inside the cover, centers this better. A light tap of the hammer will cup the cover sufficiently to supply the increase in endshake needed.

To decrease the endshake, the barrel cover is placed on the cup stump so that the inside rests on the stump. The largest solid, round-faced punch in the set is used to bump the cover. Its curvature will help to center the cover and its large size will not mar the center hole.

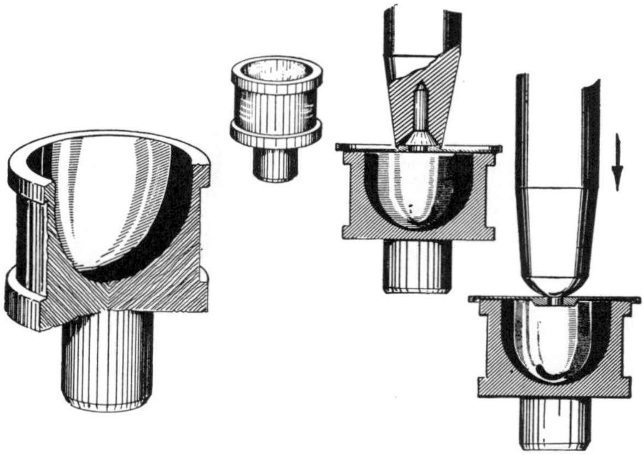

Fig. 12. Large diameter flat-faced cup stump in a cutaway version and at right, details showing how it is used to increase barrel arbor endshake (left) and decrease it (right).

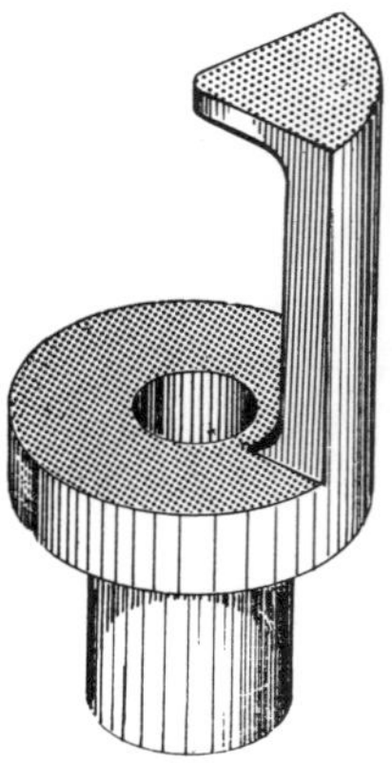

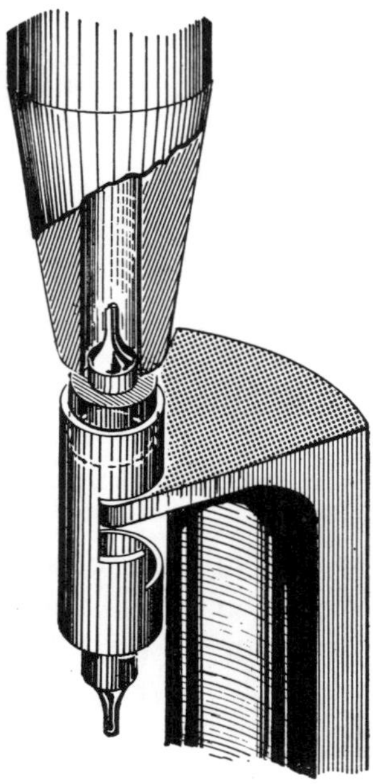

Fig. 13, left. A cylinder "cutaway" or platform stump. Fig. 14, right. A cylinder tampon or plug being driven in with the aid of the cylinder "cutaway" stump.

The stump shown in Figure 13 is used in cylinder escapement work. It is not much used in America, due in part to the unpopularity of this escapement and also in part to the lack of knowledge of American watchmakers of its adjustment. As a result, many shops refuse to accept them for repair.

The specific use of this stump is shown in Figure 14. Here the hollow cylinder is shown supported by the shelf of the stump while a (pivot) plug is being driven into the upper end of the cylinder.

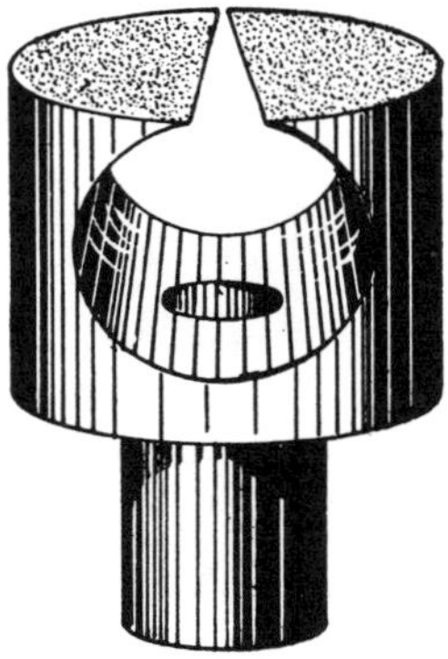

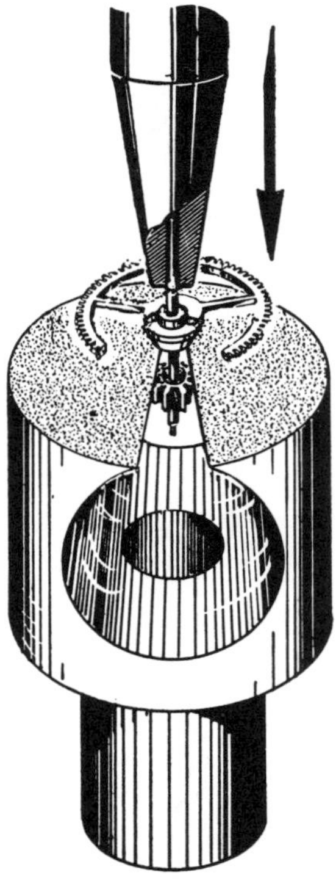

Fig. 15, left, shows a V slot stump. At right, Fig. 16 shows stump being used to aid in removing a small pinion which is set into a wheel with a bushing.

The "V" slot stump shown in Figure 15 has many uses although it was specifically designed for driving out the pinions of escape and other wheels without disturbing wheel bushings. This is especially true when the pinion is mounted higher than the wheel, the arbor is cut to a smaller diameter below it, often to provide clearance for another close wheel, and the wheel is low on the pinion, held to the pinion post by a bushing. This is shown in Figure 16. It also may be used to hold balances supplied with *Incabloc* rollers while the hairspring is being replaced. This is done by sliding the roller into the jaws of the V slot so that the jaws are between the safety and impulse rollers. This avoids crushing the weak, undercut Incabloc safety roller.

Figure 17 shows a stump with pump center. This stump has the center point as a separate piece set on a light spring. This stump serves as a base when a hole is to be closed accurately and concentric and is usually used in this operation in conjunction with a mouth-tapered punch as shown in Figure 18. When a mouth-tapered punch is used to close holes in a flat surface, there is no sure way of knowing if the mouth is concentric with the hole. This stump, used in the staking set, insures a more accurate job.

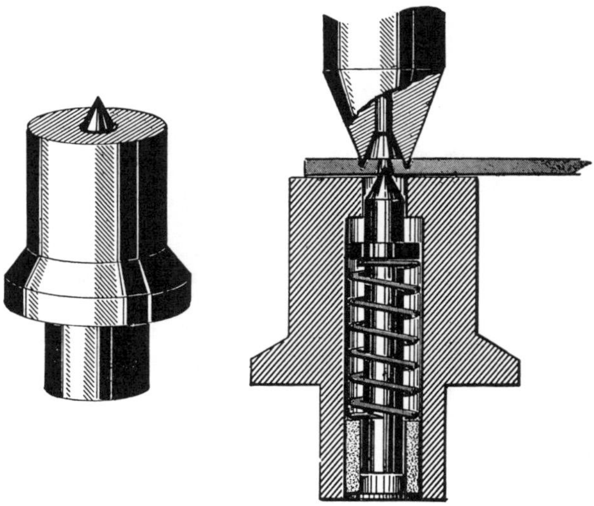

Fig. 17, left. A pump-center stump. Fig. 18, right, is a cross section view of this stump which is supporting a thin brass plate in which a hole is being concentrically closed with a taper mouth punch.

Some stumps are made to perform jobs on only one type or make of watch and come in special sets. Shown in Figure 19 is a type of stump used to hold the bevelled sides of the Waltham (hollow) balance hubs while the friction staff is being driven *out*. Here the stumps come in three sizes, 0, 12 and 16. These have holes tapered to allow the balance hub to rest partially in them while the driving out punch pushes the staff down and outward. For driving in the staff a flat-faced hollow punch or stump is employed upon which to rest the balance while specially designed punches shown here are used to drive the staff downward and in. Incidentally, these punches are especially designed so that the hub

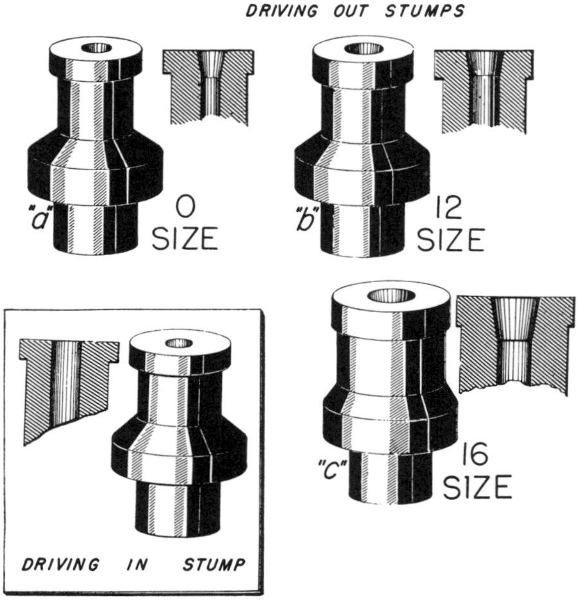

Fig. 19. Special stumps used to remove and replace friction staffs.

of the staff may be driven *below* the surface of the hub and thus be secure as well as clear. This is shown in Fig. 20. To select the proper stump in preparation for removing the Waltham friction taper shoulder staff, the blue balance hub is fitted to a stump which will accommodate the hub about ¾ of its depth. This should leave

Fig. 20. Flat-faced hollow punch is used to drive in staff. These punches are specially designed so that hub of staff may be driven below surface of the hub and thus be secure as well as clear.

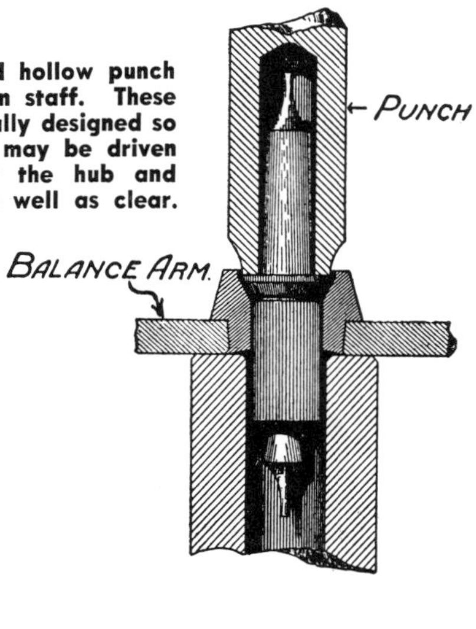

part of the hub above the surface of the stump and the balance clear of the stump's surface as shown in Figure 21.

The driving out punch which fits over the pivot and applies its force against the slopes of the coned section of the staff is then selected and a light tap of the hammer should be sufficient to eject this friction staff from the hub.

These (Waltham) stumps and punches may be purchased in a set of 4, which includes one driving-out punch for all sizes, a driving-out stump, a driving-in stump and a driving-in punch, each for 0-size, 12-size, and 16-size watches, or in all, 6 stumps and four punches.

Fig. 21. Removing Waltham friction taper shoulder staff, hub is fitted to stump which allows part of hub to remain above and balance clear of stump surface.

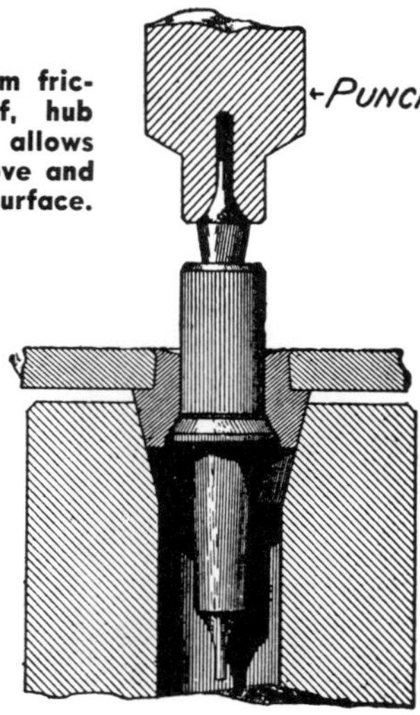

Some 23-jewel watches have escape wheel bridges in which the hole jewel, fitted into a brass setting, rests in a tight fitting recess in the escape bridge. However this jewel is held down by an elongated cap jewel which fits snug into this recess in the bridge. To further ensure that both cap and hole jewel are secure, a screw fastens the cap to the bridge.

When the screw is removed, the cap still remains fast to the bridge and will have to be removed by pressing out the brass hole jewel setting towards the cap jewel.

To give support and clearance to this odd-shaped jewel, the "U" shaped stump shown in Figure 22 is used to support the bridge while a "jewel pusher" type of punch presses upon the hole jewel setting which in turn forces the odd-shaped cap jewel down and outward.

In Figure 22, is shown an exploded view of the assembly. The escape bridge is shown upside-down in a cutaway view with the hole and cap jewel in place. Also shown is the hole jewel separately and directly below it. Directly below the hole jewel is shown the cap jewel, sometimes in this shape, other times in a figure "8" shape or a combination of both. The fastening screw is shown in the relative but disassembled position.

It can be seen how the special "U" stump will give support to this bridge while the pressure is being applied to the hole jewel setting above.

STUMPS GIVE STURDIER WORK BASE

With all this array of stumps, the watchmaker may question the advisability of their inclusion in a full staking set, especially if

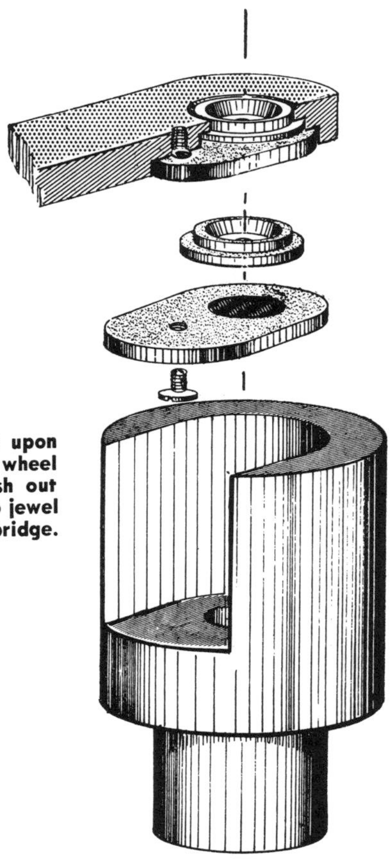

Fig. 22. Special stump upon which to rest escape wheel bridge in order to push out odd-shaped escape cap jewel set snug in recess in bridge.

the staking tool is of the type that has a base of sufficient height to allow the use of the punches as stumps in an inverted position.

If the stumps have the same surfaces and hole sizes as the punches, he may feel that this is needless duplication which raises the price of the tool. If a set has both a stump and a punch of similar surface characteristics, it is most advisable to use the stump as a base. The stump, generally, is sturdier and has a wider surface area compared to its height than the punch has. Thus if the stump is used, its shoulder rests upon the staking die-plate which was designed to take the blows to which it is subjected. The stump when hammered or struck gives a firm, solid base with little undesirable recoil, bounce or chatter.

When a punch is used as a stump, its bottom rests upon a weaker part of the staking tool, thus it does not render as firm a surface as one might expect for a riveting, swedging or indenting operation. Also when pressure is applied to a punch there is a slight but subsequent contraction of the punch through the length of the punch and the line of pressure. The longer the punch, the greater the contraction. With this contraction which is momentary, there is an expansion which results from its efforts to return to normalcy. This effect causes the "bounce" which most often is a deterrent to a successful operation. Furthermore, if the pressure is not exactly on a line with the axis of the punch, it will tend to bend under the pressure. Of course this bending is not noticeable to the eye but it is evident in the chattering or the lack of solidity in the "feel."

It should be repeated that punches are apt to become marred, chipped or worn if used needlessly as stumps.

Some watchmakers are discouraged from using the stumps because these usually become loose in the staking set boxes and eventually become lost. If manufacturers of staking sets would supply some means of firmly securing these stumps in the box so that they will stay where they are put, their use would more often find favor with the watchmaker.

BOOK III

CHAPTER I

Part I

JEWELING

JEWELS have been used more and more as bearing surfaces in timepieces since 1704, when Nicolas Facio introduced pierced rubies in watches. Now, jewel bearings are standard in all high grade movements. Quality watches are fitted with jewels ground and polished to precision dimensions.

Because jewel bearings are hard and susceptible to breakage due to stresses placed upon them, they must be replaced quite often. Jewel bearings may be set by any of 3 methods: bezel, friction, and combination jewel-in-bushing, secured to the plate by a screw.

In figure 1, "A" is the bezel set type. While this type of jewel setting is not used much in watches made in recent years, there remains on the market sufficient quantities of such watches requiring servicing and jewel replacement. Modern watches more often have jewels ground and polished so that their outside circumference is perfectly concentric with their holes. This permits them to be pressed directly into plate holes which have been accurately placed and drilled, as shown in "B" of figure 1. A jewel which has been bezel-set into a brass bushing, which in turn is pressed into the plate or held in a recess in the plate by "jewel" screws, is shown in "C," Figure 1.

Jewels serve as bearings by having holes drilled through their centers, permitting them to fit over the pivots of wheels with the square shoulder taking the end thrust ("D"); cap jewels likewise serve as bearings, taking the end thrust of the pivot.

A jewel should be replaced if it is broken, cracked, or chipped near the hole or where it may come in contact with a pivot or pivot shoulders. A cap jewel should be replaced for the same reasons, or re-ground and polished if it shows signs of pitting at the pivot end.

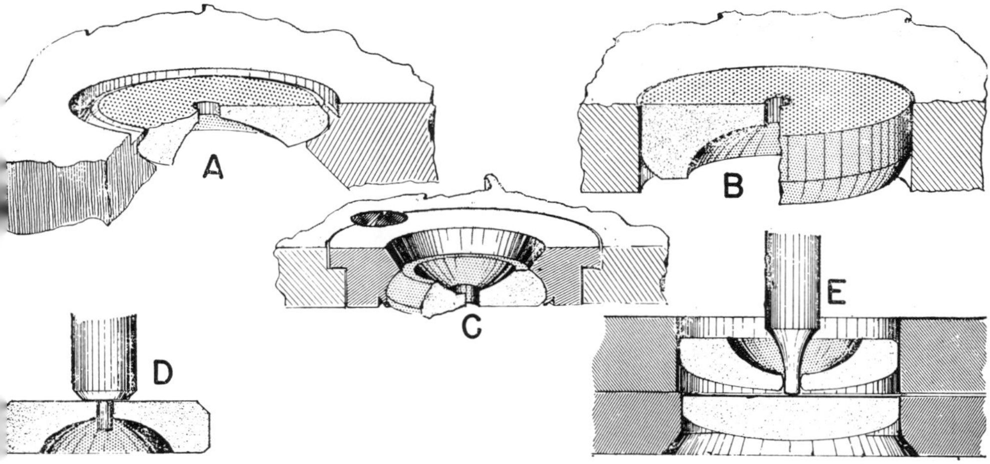

Fig. 1. Three types of jewels and the two purposes which they serve as bearings.

When replacing a jewel, determine the manner in which it was set as well as the most effective method of removing it without damage to setting or plate.

BEZEL-SET TYPE JEWELS

To replace a bezel-set jewel and preserve the setting, one should understand how it is held in place. A bezel-set jewel is similar to a bezel-set crystal in a watch case, except that the bezel over the jewel's edge has more metal overlapping it than the bezel holding a crystal. Unlike the watch crystal, the jewel is not snapped in but is burnished into the setting, and a thin-walled rim surrounding the jewel is pressed over the beveled edge of the jewel (becoming the bezel).

To replace a broken jewel satisfactorily, the setting must be restored as closely as possible to its original condition. The rubbed-over rim must be raised to permit the entry of the new jewel.

Be sure that you do not tear or damage this thin rim for it will be difficult to secure the new jewel if you do. To remove the old jewel and at the same time preserve the setting, turn a brass plug on the lathe so that the front end of the plug is flat and slightly smaller in diameter than the exposed back of the jewel. The back of this plug should have a shoulder which can fit into the hole of a flat-faced hollow punch, as shown in Figure 2. Center the plate with its jewel over a hollow stump, the hole of which is a bit larger than the full diameter of the setting. Place the brass-plugged

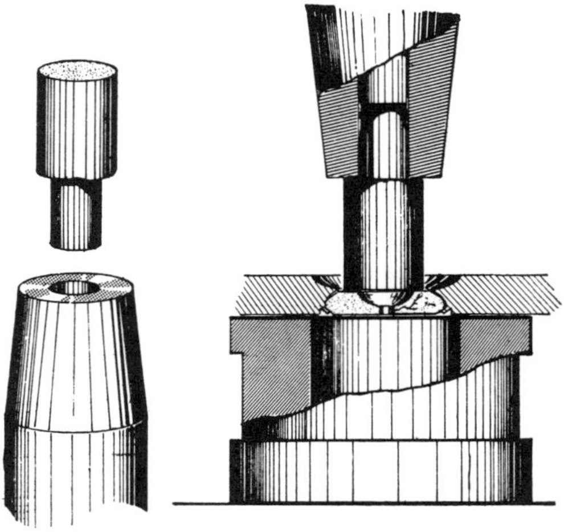

Figs. 2 and 3. Brass plug (left) to fit hollow flat-faced punch is used to push out bezel-set jewel; brass plug inserted in punch (right) is shown resting on back of jewel to be removed.

punch over the back of the jewel, as shown in Figure 3. A slight tap of the punch will not only force the jewel out but also help in raising the rubbed-over bezel, as shown in Figure 4.

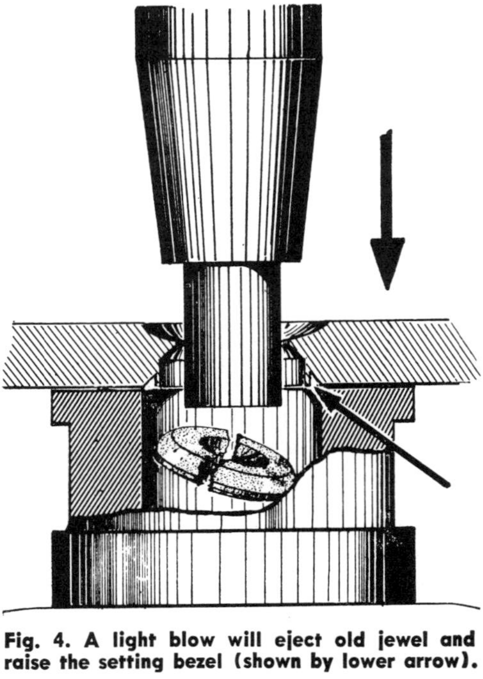

Fig. 4. A light blow will eject old jewel and raise the setting bezel (shown by lower arrow).

A friction jeweling tool also may be used to remove the old jewel. In this case, place the brass plug in the plunger, pre-adjusting the depth of the thrust so that the punch holding the plug does not come in contact with the plate or setting.

Open the bezel fully, otherwise the jewel selected will have to be smaller to fit into the partially opened setting, and will not be fully secured by the bezel after it is closed. If the bezel is not fully opened, the jewel may be set off center. The setting may be properly opened by placing the plate on a shellac chuck or on a face plate and centered; then burnish the bezel of the setting outward and open. The setting also may be properly opened by using special bezel openers. (In some cases the bezel may be opened by rubbing it outward with a pointed burnisher held in the hand; however, this method is not recommended because it requires great skill and there always is danger of slipping and tearing the thin bezel.)

SHELLAC CHUCKS

The method of shellacking a watch plate to a brass plate held in

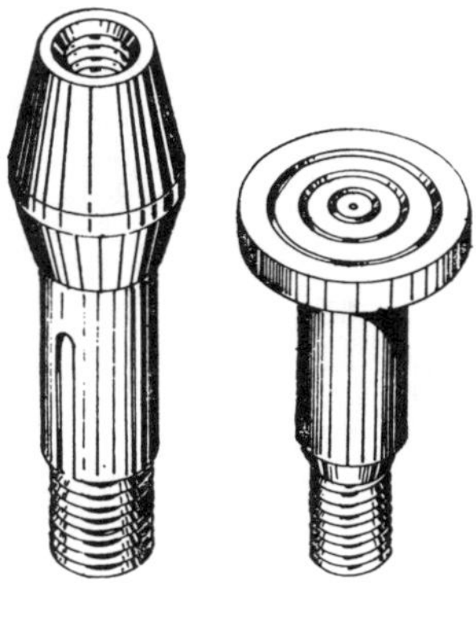

Fig. 5. Shellac (cement) chuck of brass screws into the lathe chuck; circular grooves provide space for shellac.

the lathe chuck is relatively simple and reliable. Clear the broken jewel from its setting and clean out the chips of jewel by running a pointed pegwood into the setting. Heat the brass plate (shown in Figure 5) and cover its face with a thin layer of shellac. The watch plate is heated moderately and placed on the brass plate, with its setting facing outward and as close to the center of the chuck as the eye can judge. The "T" rest is adjusted so that it is close to the plate and raised almost to the height of the jewel. The edge of the "T" rest should be turned parallel with the plate and at right angles to its axis.

To center the watch plate perfectly, the lathe is turned *clockwise* at moderate speed. While the lathe is in motion, heat the assembly (watch plate and the brass plate in the chuck) by an alcohol lamp (held in the left hand); and hold a pointed piece of pegwood in the right hand (rested on the "T"-rest) against the setting with the edge of the pegwood point touching the inside circumference of the setting, causing the plate and setting to shift true to center. When there is no resistance to the pegwood and the setting is centered satisfactorily, remove the alcohol lamp and keep the lathe spinning with the pegwood guiding the setting until the shellac sets and hardens. This operation is shown in Figure 6. The centered jewel bezel may be opened, and the new jewel inserted and set.

JEWEL SETTING (BEZEL) OPENERS

When the watch plate is in the lathe, the bezel may be opened by

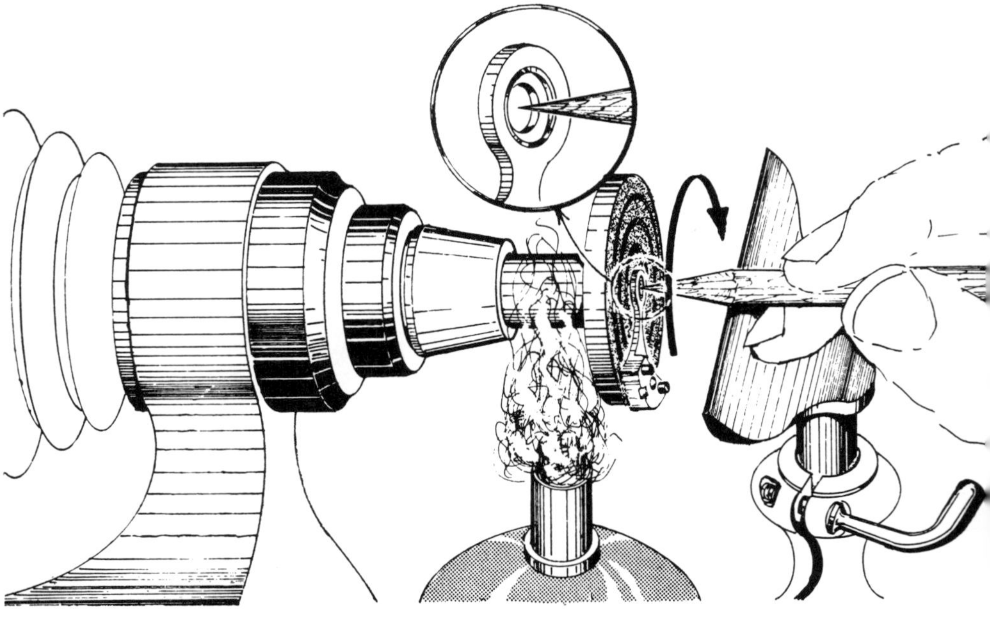

Fig. 6. How work held in a shellac chuck may be perfectly centered, using heat-softened shellac, peg-wood, and T-rest.

placing a pointed (but not sharp) polished hard steel burnisher inside the bezel; while the lathe spins, apply this burnisher outward with a little pressure. This causes the bezel to open. A drop of thin oil on the burnisher will prevent it from "grabbing" and tearing the thin metal of the bezel.

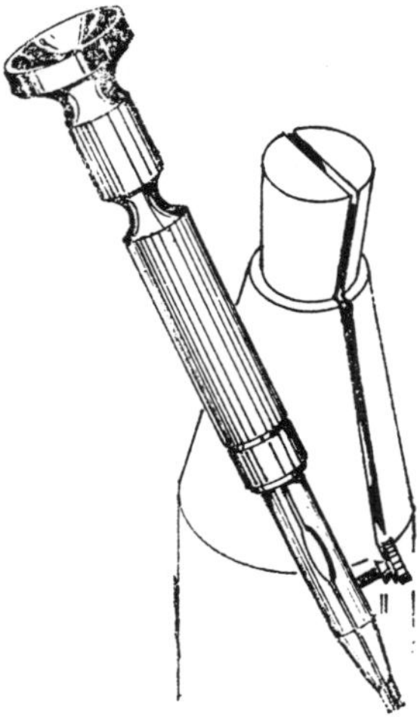

Fig. 7. A jewel bezel opener, used when bezel is opened without lathe work.

To open the setting (after the jewel is removed) without use of the lathe, jewel setting openers may be used (shown in Figure 9.) The jewel setting openers have 2 slightly undercut jaws adjusted to fit into the empty bezel, and an adjusting screw which is opened so that a slight pressure is felt against the bezel. Twisting the

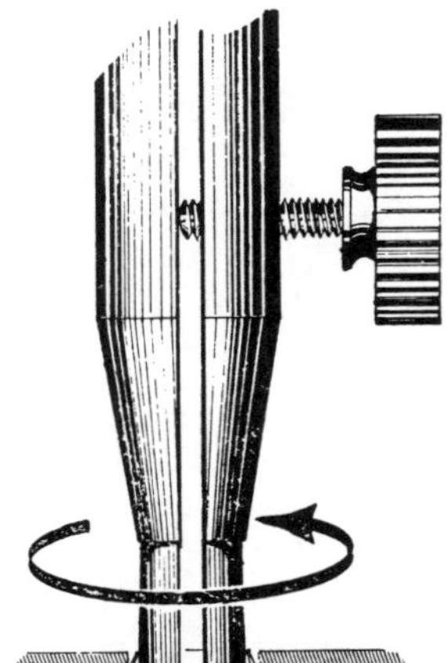

Fig. 8. How bezel openers fit into empty jewel bezel, and twisted to raise the setting.

tool forces the bezel outward. The adjusting screw is opened to permit the jaws to spread a little more, and the operation (shown in Figure 8) is repeated until the bezel is fully opened.

The new jewel should duplicate the hole size, diameter, thickness, and general shape of the jewel it replaces. If the old jewel is broken or shattered, the hole size may be obtained from the pivot of the wheel which is to enter the new jewel. To obtain the diameter of the jewel, gauge the diameter of the bezel opener when it was last opened to fit into the jewel setting.

When a plate jewel is set, its flat top surface usually is flush with the plate. The thickness of the jewel selected must conform to these requirements, otherwise the endshake will be improper.

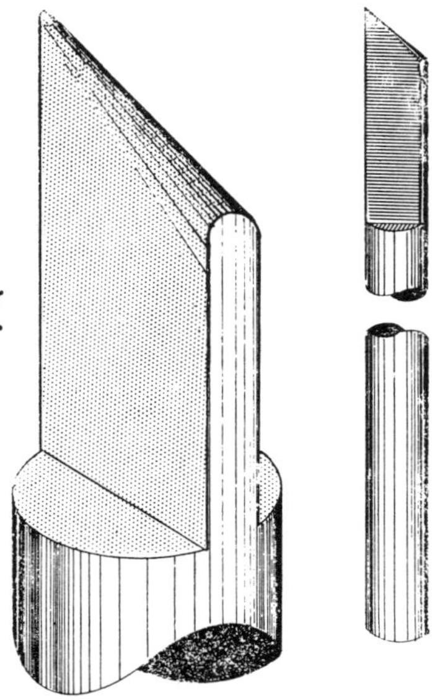

Fig. 9. Two views of a popular type of jewel bezel burnisher.

When a correct replacement jewel has been selected, place it loosely in its setting, using a little watch oil to prevent the jewel from falling out. Burnish the bezel over the bevelled edge of the jewel.

BURNISHING JEWEL IN SETTING

Watchmakers with jeweling experience have a variety of large steel knitting needles which they have converted into jewel setting burnishers. Some watchmakers use phonograph needles, slightly blunting the r tips to prevent cutting into any part of the setting. One of the most popular shapes of jewel bezel burnishers is shown in Figure 9. Its angle and shape is ideally suited for spinning over the bezel with the least effort and without danger of tearing it. This burnisher is made from high carbon steel (drill rod) about 2.50 mm. thick and about 5 inches long. It is shaped, smoothed with emery, hardened, and polished with crocus cloth or diamantine on a boxwood block.

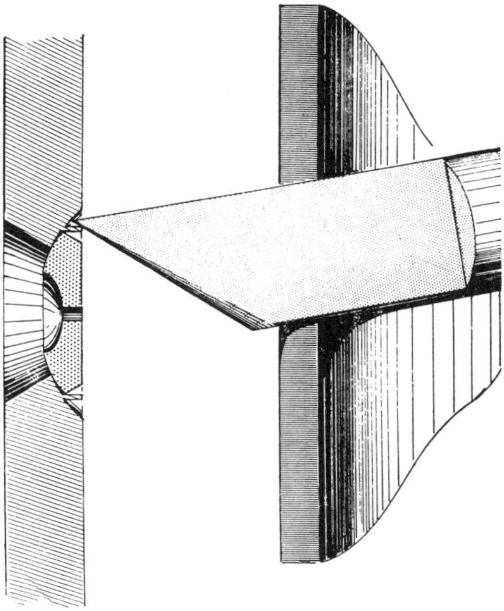

Fig. 10. How this type of burnisher is applied to the bezel groove.

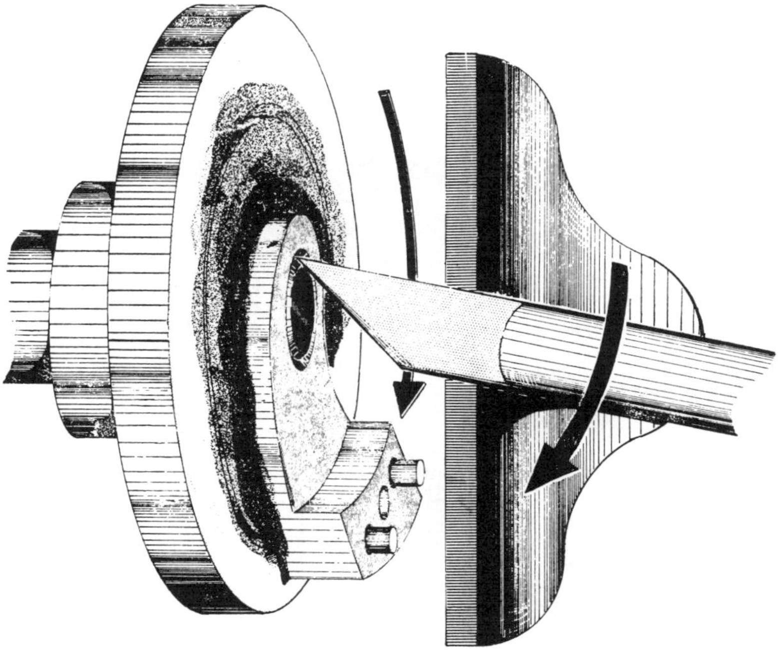

Fig. 11. Showing the shellacked, centered bridge having its jewel setting burnished over, while the lathe turns in the direction shown by the arrow; move burnisher slowly as indicated by arrow.

To secure the jewel in its setting (as shown in Figure 10), apply the burnisher to the groove outside the bezel on the far side of the setting while the lathe is in motion, turning in a clockwise direction (facing the chuck). A drop of clean oil applied to the tip of the burnisher will prevent "grabbing" of the metal. The T-rest, as shown in Figure 10, is placed at right angles to the lathe's axis and raised to permit the burnisher to contact the work readily and give it support. With the lathe in motion, ease the lubricated burnisher gently into position, apply a little pressure, and move it slowly in an arc, as shown by the arrow in Figure 11. The burnisher thus presses the metal over the bevel of the jewel, centering and holding it securely in place.

MAKING A JEWEL SETTING

Sometimes it is necessary to set a jewel into a brass (bushing) setting, which in turn is fitted into a recess or counterbore in the plate or bridge such as is found in American and some Swiss balance bridges. This arrangement has certain advantages over the bezel

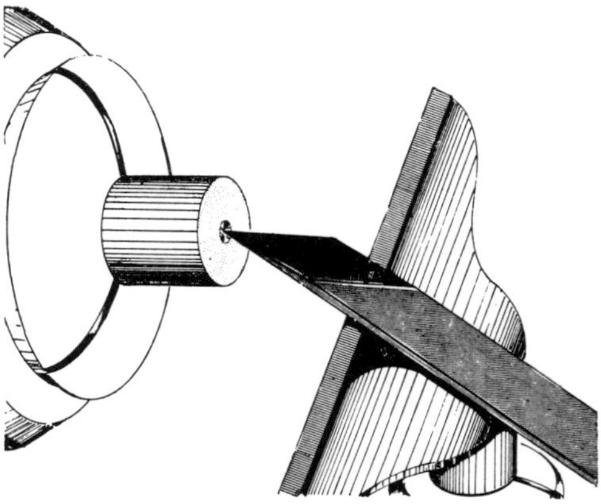

Fig. 12. The first operation in making a jewel setting is to provide a center guide.

set jewel placed directly into a plate or bridge. Many jewels can be replaced on one wheel pivot position without seriously altering the center distance between wheels. If a jewel in a bushing has an off-center hole, the jewel hole may be centered on a shellacked chuck, and the bushing trimmed concentric with the jewel hole.

To employ the foregoing arrangement, suppose that the sample is lost and measurements must be taken from its position in the movement (in this case from the balance bridge).

Select a balance jewel that fits the balance pivot accurately. Setting the balance jewel into a brass rod which, when shaped in the lathe, becomes the setting to fit the balance bridge. Use a length of rod brass about 4 mm. thick, holding it in the lathe by a close fitting chuck. Apply a sharp graver carefully and gently to the face end of the metal, and cut a slight countersink at the center of the metal to guide a drill point, as shown in Figure 12.

Using a drill about three-fourths the diameter of the jewel, bore a hole about 6.00 mm. deep (Figure 13).

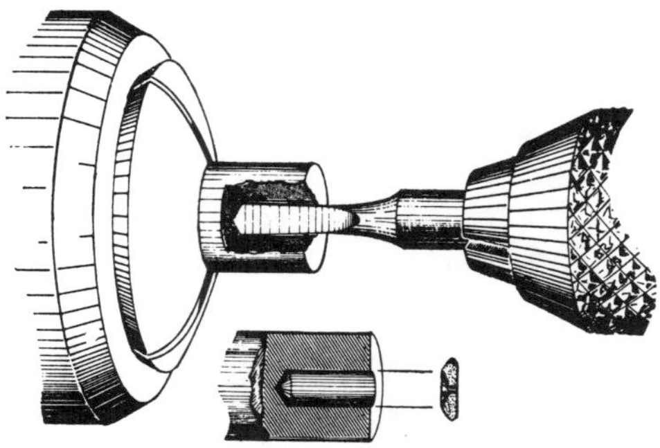

Fig. 13. A hole about 3/4 the diameter of the jewel is drilled to a depth of approximately 6.00 mm.

Make a seating for the balance jewel. The surface of the balance jewel (which will contact the seating) probably may be curved; if it is curved, make a seat to conform to this shape. The depth of this seat should be .02 mm. deeper than the thickness of the jewel, and the width of the seat should be the width of the jewel. A boring tool shown in Figure 14, may be made of drill rod, hardened,

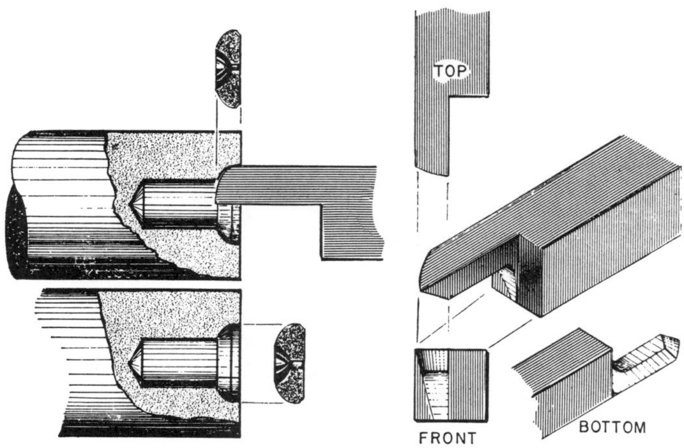

Fig. 14. Details of a boring tool which can cut the seating for the jewel.

and tempered to a light straw color. (A specially-shaped graver also may be used for boring, but in this case be exceedingly careful in cutting the seat so that you do not cut the edge of the countersink).

Many watchmakers test the depth of the counterbore with the jewel. In the case of a plate jewel, many test it by placing the jewel on the flat end of a pegwood stick held by a spot of beeswax. Try the plate jewel into the seating, and if it is below the edge of the counterbore or countersink, face the surface of the bushing

Fig. 15. For a balance jewel setting, top of setting should have clearance indicated between top of jewel and of setting

until the jewel is flush. If the plate jewel does not sit deep enough, deepen the seating. When the jewel is flush, cut the thin rimmed bezel. (If the setting is to be for a balance jewel, the top of the jewel must be slightly below the top of its bushing for a clearance of about .02 mm. between it and the surface of the cap jewel for oil, as shown in Figure 15).

Watchmakers experienced in setting jewels often prefer to cut the thin rimmed bezel while the jewel is in its seating; on the other

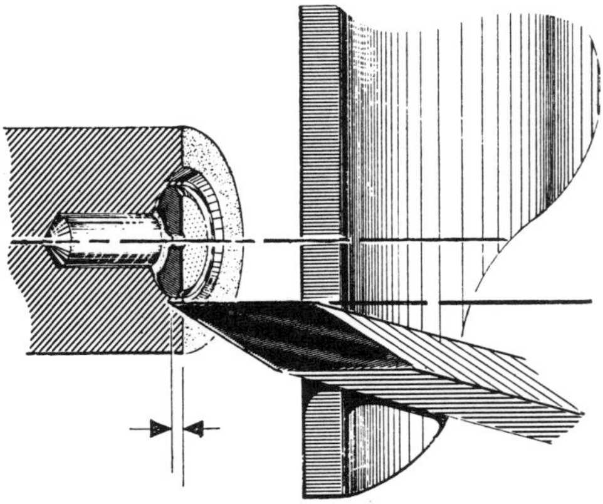

Fig. 16. Appyling a sharp graver to cut the bezel: notice that the cutting edge is held parallel to lathe axis.

hand without anything inside the hollow counterbore, the thin-walled rim may be bent over as the graver cuts it, reducing the entrance diameter of the counterbore making it difficult for the jewel to enter its setting. Oil the jewel to hold it in place while the rim is being cut (or use a pinhead of beeswax), as shown in

Figure 16. While being cut, the rim will bend slightly away from the graver and slightly over the jewel, holding it securely for the final burnishing-over operation.

Finally, to cut the thin wall rim, apply the graver to the face of the metal close to the edge of the counterbore in which the jewel is seated. (It is advisable to cut this rim thin so that it can be spun over without pressing too hard on the rim and without any loss of "feel" or danger of cracking the jewel. The graver recommended for this purpose is one with an acute angle, applied to the setting as shown in Figure 16.

The T-rest is still in the same position, facing the front of the stock. The groove should be as deep as the *bevel* of the jewel. This is shown by the dimensional arrows in Figure 16.

In making jewel settings, watchmakers occasionally make a set of jewel setting cutters for cutting the counterbore and the bezel. If a set is made, make it so the cutters fit into the tailstock spindle. If some round stock steel is on hand, cut it carefully to the required angle to fit the taper in the spindle, using a slide-rest set to the correct angle when cutting. If steel with unfinished ends is bought, shape them to the desired form and dimensions.

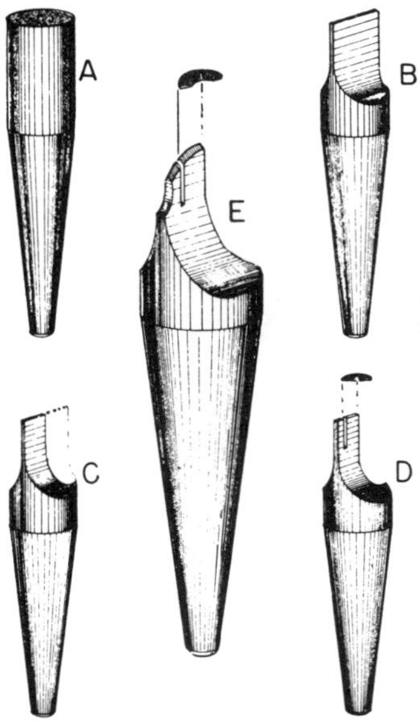

Fig. 17. Tailstock-taper jewel setting cutter: taper (A), grinding to chisel shape (B), half of taper ground away (C), thin slot cut to depth of jewel's radius (D), cutter face curved (desired slope of jewel set), sharpened (E).

Steps in making a cutter are shown in Figure 17. The unfinished end of the cutter stock is filed or ground to a screwdriver shape (B). In this screwdriver shape the cutter would cut a counterbore the diameter of the tool; however, if the tool had half of its side removed, it still would cut a countersink of the same diameter and, in fact, would make a cleaner cut (C). Use a fine jeweler's saw to cut a slot into the edge of the cutter (D). Shape the end of the cutter (E).

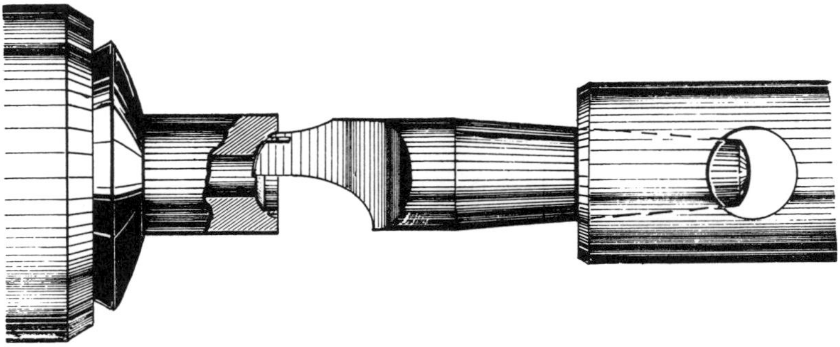

Fig. 18. Tailstock-taper seating cutter makes a seating and bezel in one operation.

To cut with the bright finish desired in good work, polish the cutting surfaces of the tool after hardening and tempering. The cutter will leave a thin-walled rim between the outside groove and the countersink, which becomes the bezel (Figure 18). A set of 10

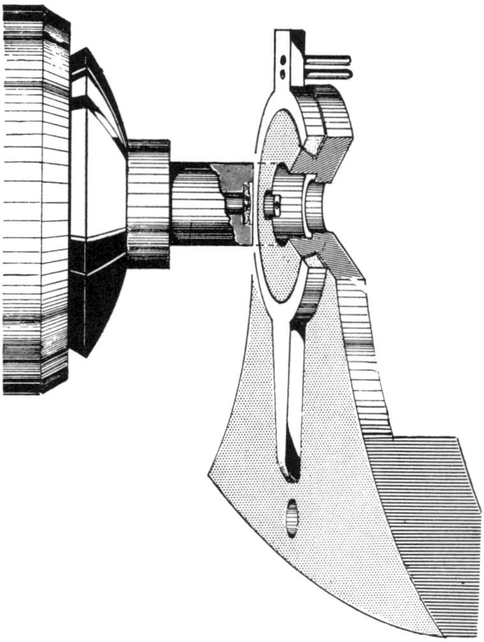

Fig. 19. Jeweled rod is cut to fit into upper recess of balance bridge.

cutters in various widths usually is sufficient to cover most jewel sizes.

Although bezel burnishers come in many styles, the one shown in Figure 9 will do for most work. Spin or burnish the thin-walled rim over the beveled edge of the jewel, as explained in earlier paragraphs (see Figures 10 and 11); do not press too hard, and use a little lubricant and keep the burnisher highly polished and free from scratches.

When the jewel is secure in its setting, trim the outside diameter of the bushing to fit snugly into the stepped opening of the balance bridge (Figure 19). The wire now holding the jewel bushing may be turned to this new diameter and tested so that it fits snugly into the larger upper part of the stepped opening. Separate the brass bushing with its jewel from the remainder of the rod at least the thickness of the balance bridge at the end in which the jewels are housed.

After the thin bushing with its jewel is separated from the rod, cut the lower smaller shoulder holding the bushing and jewel in the lathe.

STEPPING DEVICES

The bushing which must be extended from the lathe chuck for cutting often is not long enough to leave sufficient metal remaining in the chuck to permit it to be held firmly, let alone accurately.

A few methods and tools allow a shallow yet accurate and firm grip on the jewel bushing for the "stepping" operation (making a

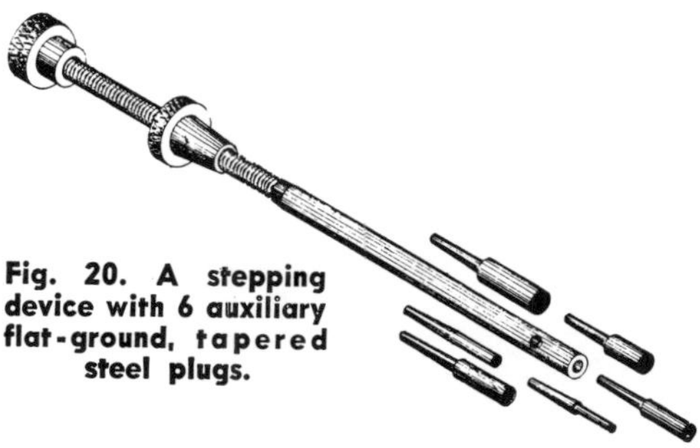

Fig. 20. A stepping device with 6 auxiliary flat-ground, tapered steel plugs.

shoulder). These tools are called *stepping devices* and *jewel chucks;* both work on similar principles. A stepping device has a number of steel plugs of various diameters matching the jewel bushing to be stepped (**Figure 20**).

A stepping device, used in conjunction with the regular lathe chucks, is shown in Figure 21. Place in the lathe a chuck of the size of the jewel bushing. Insert the stepping device through the rear of the lathe head into the draw-in spindle. Wedge the tapered rear nut of the stepping device into the hollow spindle bar. Turn

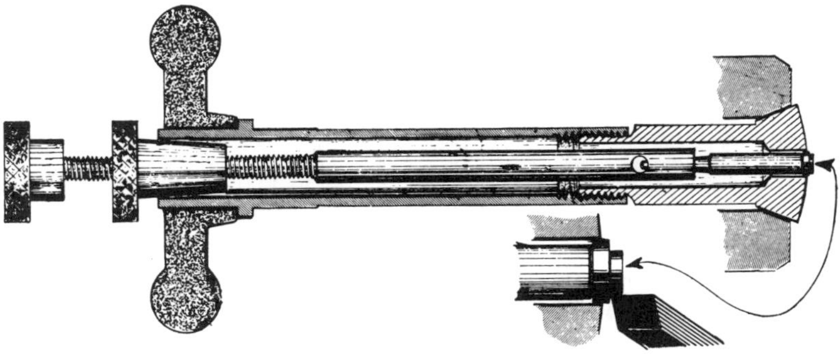

Fig. 21. Stepping device fastened through the draw-in-spindle, and supporting the jewel setting.

the knob of the stepping device until the front plug entering the front split hole of the chuck contacts the bushing and pushes it out of the chuck sufficiently so that the step or shoulder may be cut. The advantage of the foregoing procedure is that the bushing with

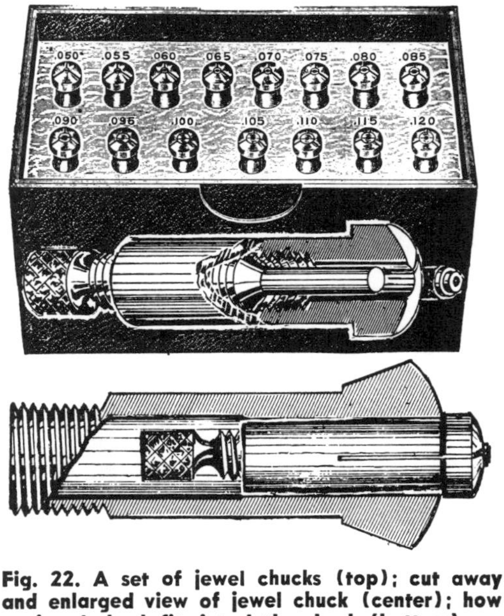

Fig. 22. A set of jewel chucks (top); cut away and enlarged view of jewel chuck (center); how jewel chuck fits into lathe chuck (bottom).

its jewel may be removed from the chuck for trial fitting as often as may be required. When returning the bushing to the chuck for any further alteration, place it on the finger-tip with the jewel-side up and into the chuck. The front plug of the stepping device prevents the bushing from going too far into the chuck, and the precision-ground face of the plug makes a true seating for the back of

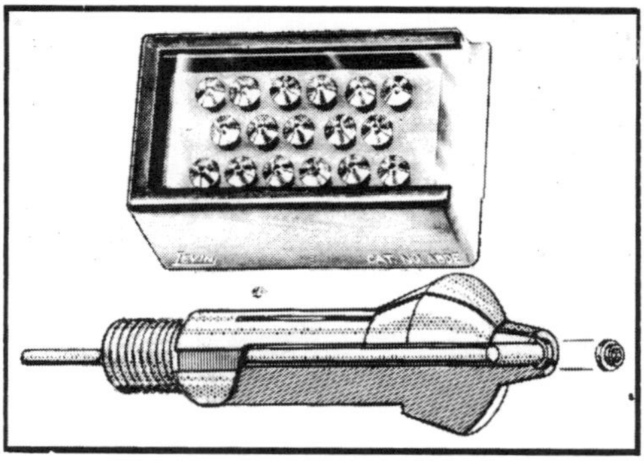

Fig. 23. A set of special lathe chucks designed expressly for making and adjusting jewel settings.

the jewel setting—and the jewel will lie flat in the bridge after cutting the shoulders.

JEWEL CHUCKS

Jewel chucks are graduated in size, each fitted with its own adjustable plug stop (Figure 22). They are 3/16″ in outside diameter and will fit into a No. 48 lathe chuck, although some sets are designed to take a No. 50 chuck (5 mm).

Place the jeweled setting into the chuck, the opening receiving the bushing. Adjust the screwplug to permit the bushing to sit as shallow as required. Tighten the regular lathe chuck, contracting the jaws of the jewel chuck around the bushing. While these chucks are easy to work with and in principle are good, nevertheless because a jewel setting must be inserted into a chuck of questionable accuracy, and in turn, both into another chuck, may compound the errors and may result in a setting which is off center.

LATHE CHUCKS

A set of regular lathe chucks, designed to eliminate errors inherent in other stepping devices and jewel chucks (Figure 23), has tapered noses designed for accessibility of the cutting graver. The tips of the chucks are annealed so that they may be readily cut to fit a special setting, should the occasion arise. Plug stops of each chuck prevent collapse of the setting, should the setting be necessarily thin and the pressure upon the chuck endanger its shape.

A method of holding the jewel setting without the use of special (jewel) chucks is by cementing the jewel bushing to a piece of brass rod, which should be a little thinner than the bushing, offers another accurate method of cutting the step and maintaining a

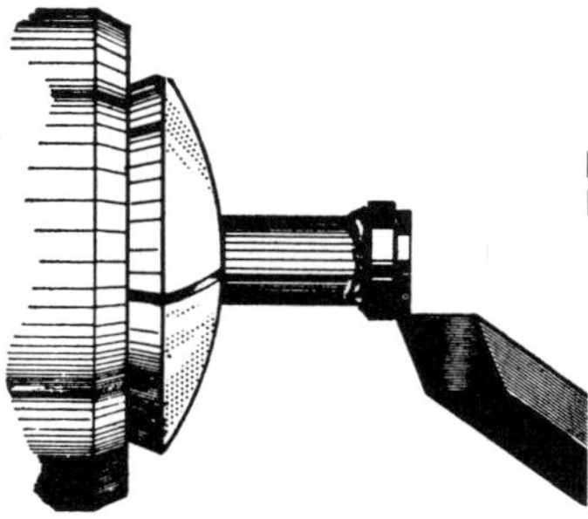

Fig. 24. Setting may also be shellacked to a brass rod, centered, and "stepped" or cut.

shoulder whose circumference is concentric with the jewel hole (Figure 24). Cementing and centering the bushing on the brass wire follows the same procedure as illustrated in Figure 6, except that the cement brass is thinner than the setting, thus facilitating cutting of the jewel setting shoulder by allowing greater accesibility to the graver.

In making a balance jewel setting without the sample, stepping devices or jewel chucks are preferred because the operation can be performed with the least effort, and futhermore the setting may be removed occasionally for trial fitting. When trial-fitting the balance jewel setting, its shoulder should be high enough to allow the setting to sit into the balance bridge so that the balance pivot reaches just to the top of the balance jewel hole. If the pivot does not extend to the top of the hole during the trial fit, return the setting to the lathe chuck and make the step higher, allowing the setting to go lower into the balance bridge and the staff pivot to reach further into the jewel hole.

Should the pivot extend through the jewel hole, the position of the pivot indicates that the step is too high, requiring the jewel to be reset into another bushing and restepped.

If the jewel setting has been cut to fit the balance bridge and the steps allow for proper entry of the balance pivot and for endshake the setting is ready to be finished. If the step of the setting extends through the balance bridge, remove the protruding metal as it may intrude upon the hairspring collet. The step should be faced off until it is flush with the under-side of the balance bridge. The setting is finished by bevelling out the metal around the jewel's oil sink in front of the smaller shoulder.

Since this lower part of the setting will be exposed to view when it is fitted to the bridge, it should present a pleasing, workmanship-like appearance. This bevel should have either a curved-out bevel cut or a straight bevel cut with a bright finish. This finish is obtained by taking a light final cut with a "stripping" graver.

Part II

UPRIGHTING

STRIPPING gravers, when applied to the setting turning in the lathe, remove a fine layer from the surface, leaving a shiny, bright finish. Stripping gravers of hard steel have the cutting face polished to a high black finish. Other stripping gravers are polished carboloy and special shaped bits of sharpened and polished sapphire. Many watchmakers prefer steel stripping gravers which work as well as the others, because they are inexpensive, can easily be sharpened without the need of special equipment (diamond lap) and can be reshaped if desired.

To convert a regular lathe graver into a stripping graver sharpen the surface by using a hard Arkansas oilstone, clean it thoroughly, and pull it across a clean boxwood block upon which some dry *diamantine* has been spread. Successive one-way pulling strokes of the graver across the block produce the desired finish as shown in Figure 25.

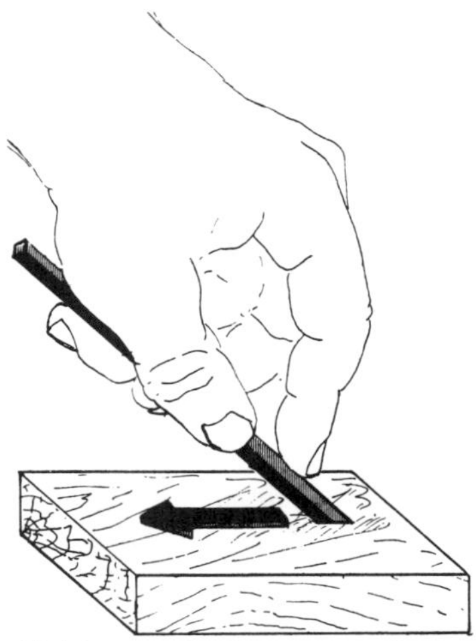

Fig. 25. Finish a stripping graver by stroking it across a wooden block spread with dry diamantine.

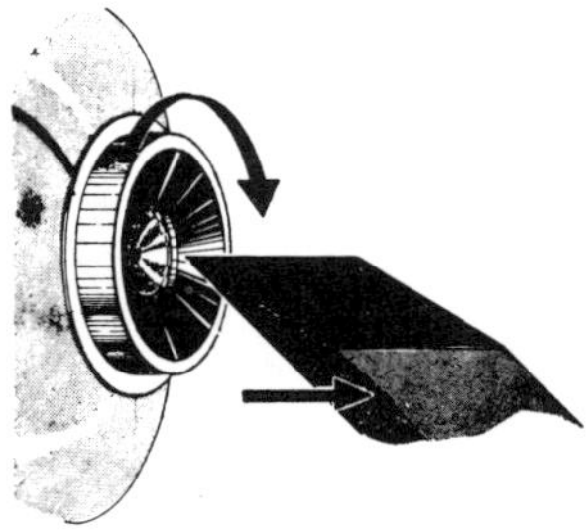

Fig. 26. Move the graver in and out of the bevel to give a bright cut as shown above.

In applying the graver to the bevel of the setting, run the lathe at medium speed, the graver taking light cuts and moving slowly from the center to the outside, as shown in Figure 26. The shaping of the bevel should not encroach upon the edge of the step and

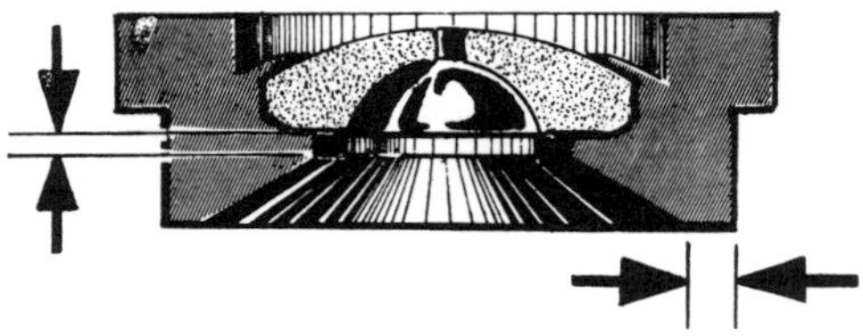

Fig. 27. The bevel should be cut so that the rim-like edges (arrows) are retained.

should not touch the jewel, but as a safeguard should leave slight flat surfaces, as shown by the arrows in Figure 27.

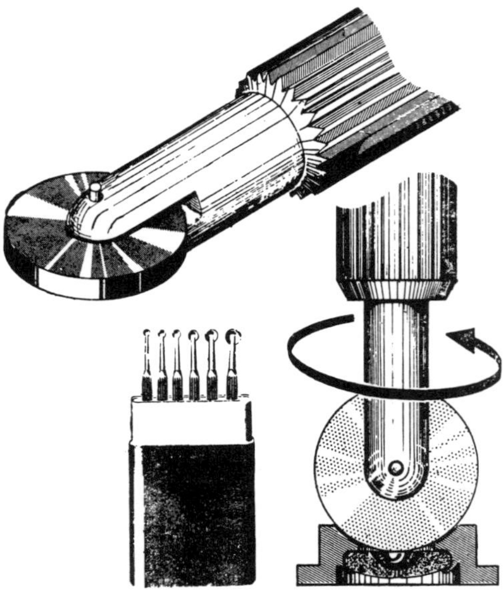

Fig. 28. A set of disc-countersinks and how they are used to countersink the bevel of a jewel setting.

Sometimes it is desirable to make a curved type of inside bevel. This is made with a special graver or with a set of countersinks (shown in Figure 28). These countersinks are small, hard, polished steel discs pivoted in the center and set as rollers into handles. The sharp edges of these countersinks can cut polished, curved countersinks in settings, and they also can deburr holes after broaching, filing, or punching.

MAKING A PLATE JEWEL SETTING

To make a plate jewel setting, stepping devices or jewel chucks,

Fig. 29. When plate jewels are held in settings it appears as at left, with the depth of the step governing the endshake.

stripping gravers, and countersinks are again used. The flat top of the hole jewel which contacts the pivot and its shoulder is usually set face up on the smaller of the 2 shoulders of the setting, as shown in Figure 29.

In this case, the amount of endshake that is assigned to these pivots is governed by the height of the step. Also the countersink is seen from the movement side and also must be polished.

Fig. 30. Cap and hole jewels which are held in settings have these settings held in place by screws set into countersinks.

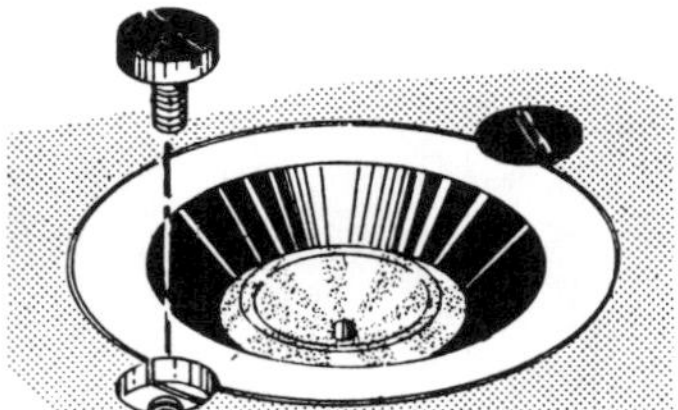

Often, in the case of set plate jewels, these are held down by screws. These screws are seated flush in countersinks in the plate or bridge and overlap the edge of the jewel setting in a partial countersink on the edge of the setting as shown in Figure 30. In this case, countersinks at the edge of the setting will have to be provided after the setting has been made. This is done by placing the setting in position in the plate or bridge. Then a countersinking cutter from a set, such as shown in Figure 31 is chosen which most closely fits the screwhead countersink in the plate. This cutter should have a pilot pin or pivot (shown by the arrow) which fits into the threaded section of the screwhead countersink in the

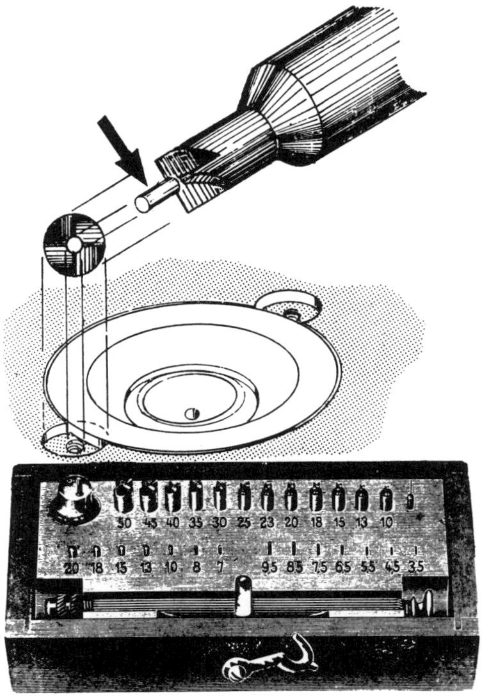

Fig. 31. Showing a set of counterboring cutters and how one is selected to cut the countersink in the setting's edge.

plate and it guides the cutter in forming the countersink without drifting. The cutter will only shave off the metal on the edge of the jewel setting since the screwhead countersink in the plate is already formed. The countersink in the edge of the setting should be no deeper than the countersink in the plate (Fig. 31).

If no countersinking tool is available, a suitable one can be made by following the steps shown in Figure 32. The pilot pin is turned down to the diameter of the threaded hole as in ("A"). The next operation is cutting the shoulder of the stock on which the cutting blades are made. This is cut until its diameter is the same as that of the countersink in the plate which accommodates the jewel screw ("B").

The cutting edges in this shoulder may be ground or milled with a milling attachment in the lathe or it may be filed ("C"). File into the edge of the shoulder with upward strokes. Then the stock is turned one quarter way and then filed again, after which the

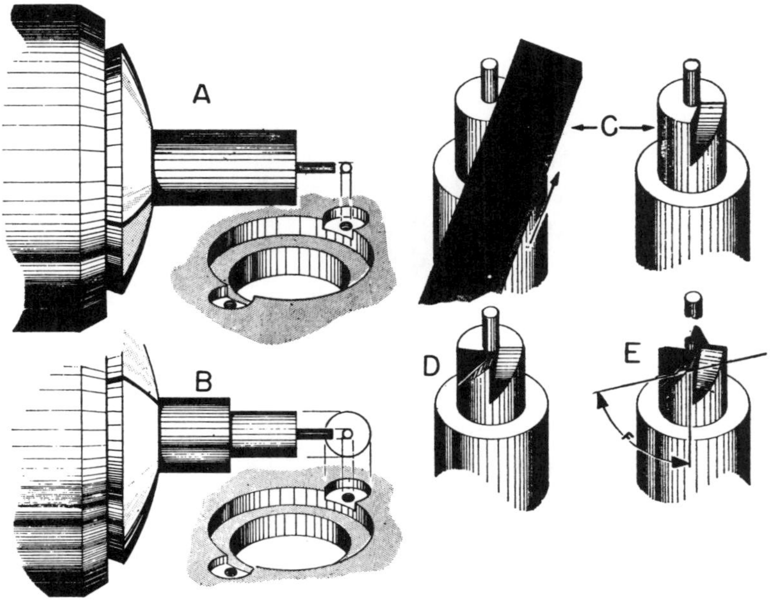

Fig. 32. Making a countersinking tool: the pilot pin is turned down to the diameter of the threaded hole ("A"); and the shoulder of the stock, on which the cutting blades are made, is cut to the diameter of the countersink in the plate which accommodates the jewel ("B"). The tool is then ground, milled ("C"), or filed ("D," "E," "F").

countersinking tool should appear as in ("D.") After filing the remaining 2 quarter edges, the tool should appear as in ("E.") The cutting angle ("F") is filed or stoned to give a stronger cutting edge to the countersink. Harden and temper the tool and set it into a screwdriver-like handle. This countersinking tool will mill countersinks as well as any factory-made tool.

POLISHING THE TOPS OF JEWEL SETTINGS

Polishing this part of a jewel setting may be done by using polishing compounds or by burnishing. To polish by the use of compounds, the setting is placed top-side-down upon a piece of fine frosted glass onto which a small bit of diamantine has been placed. The setting is then rubbed across the surface of the glass until the fine flat polish is obtained. When a boxwood block is used, the diamantine applied to the wooden surface should be dry. Another method which produces excellent results is with the use of a large slab of flat jasper (stone). The setting is rubbed across the stone's surface and the natural elements of the stone will bring up a bright finish.

BURNISHING

The tops of the jewel settings may be polished by rubbing them across a flat burnisher. The burnisher in this case may be the plain back of a file which has a filing surface on one side only. The back of the file is then dressed with a coarse grade emery stick with the grains or marks of the emery stick going *across* the burnisher. Then the burnisher is thoroughly cleaned with benzine, making certain that the fingers do not come in contact with the burnishing surface.

The jewel setting is then ground flat by rubbing it across a ground glass surface prepared with oilstone powder. The setting is then cleaned with equal thoroughness and dried, again making sure that the fingers do not contact the surface to be polished. The setting is then placed upon the burnisher with the tweezers and the finger is placed upon the setting, separated by a piece of tissue paper from the setting. The setting is rubbed along the length of the burnisher, across the emeried grains on the burnisher, using considerable pressure on the setting. This will soon bring up a high burnished surface to the setting. During this entire operation of polishing the top surface of the setting, care must be exercised that the jewel setting is not *ground* down so much that when it is fitted to the movement, its top will be below the surface of the movement. Therefore when first cutting the setting, some allowance should be made for this (later) operation of polishing by leaving the top just a little higher.

REPAIRING DAMAGED SETTINGS

If a jewel is set into a damaged setting, it may upset the center-distances. There are a few methods of repairing damaged settings and restoring the center-distances. If the bezel is damaged but the seating for the jewel has not been marred, the plate may be cemented to the shellac chuck and centered by the method illustrated in Figure 6. The pegwood centering the plate is guided by the undamaged jewel *seating*. The setting may then be enlarged and a new bezel cut to accommodate the new but larger diameter jewel. Or, the hole may be enlarged and a jewelled bushing inserted. Another method is to enlarge the hole and to insert a friction jewel (Figure 1-B) by the methods which will be described a little later in this chapter.

UPRIGHTING

When a jewel setting or bushing is completely damaged and the center-distances obviously distorted, these distances may be restored by centering up the opposite, undamaged plate or bridge in the lathe, with the damaged plate or bridge attached. This operation is called "UPRIGHTING."

Uprighting may be accomplished by using a universal face plate in the lathe or a similar device in conjunction with a friction jewelling tool. Uprighting by cementing may be done by centering the opposite, undamaged setting to the cement chuck and then assembling the damaged bridge, after which a new hole is cut which will be concentric with the opposite hole. Then a new jewel setting is cut to it, or a jewelled bushing may be fitted to the centered-out hole.

HOW TO USE A UNIVERSAL FACE PLATE

The universal face plate, which is an attachment to the headstock of the lathe, is used for uprighting plates and bridges, milling recesses in plates and bridges, and to make plates and bridges. In spite of its versatility, the universal face plate is seldom employed by watchmakers because instruction for its use is not readily available. (The face plate is shown in an exploded view in Figure 33).

The hardened, grooved steel center-point ("A") fits through the top of the chuck and is guided by a pin in the chuck; the front end of the adjusting rod ("B") is threaded into the rear of the steel center and manipulated to move the center-point in or out (this is called the "pump-center"); the face plate ("C") is nearly 4″ in diameter; peepholes ("D") furnish "windows" permitting pointed center "A" to be observed while the plate or bridge held in the face plate is guided to the center-point; the face plate is mounted on the lathe chuck ("E") and is secured by three screws ("F"); one of three doghead clamps is shown in three different views; in ("G") the clamp is attached to the face plate and holds a bridge; ("H") is another clamp fully closed; ("J") is the top clamp-jaw of the third clamp shown with back screw ("L") and stop nut ("M"); ("K") is the lower clamp jaw.

Post ("N") fits through the lower clamp hole ("O"); the tension spiral spring ("Q") fits into the special recess ("R") supplying the opening power when the nut ("S") with its washer ("T") is loosened; threaded post ("N") fits through the lower clamp and through shift slots ("U"), into the washer ("T"), and is fastened by knurled nut ("S"); threaded post ("L"), friction

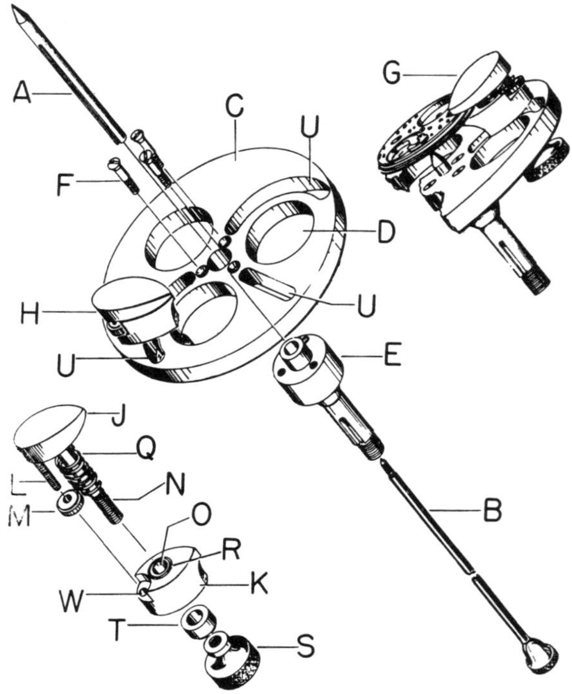

Fig. 33. Enlarged detail of face plate chuck. Some face plates come permanently attached to a special head stock.

tight in the top of clamp ("J"), goes into unthreaded hole ("W") in the lower clamp ("K").

The nut ("M") which fits on post ("L") is elevated or lowered to regulate the distance between the upper and lower jaws, insuring that the grip on a plate or bridge by these two jaws is parallel, much like the rear bolt of a toolmaker's clamp (parallel clamp). The nut ("M") also serves to take some of the strain off post ("N") when the jaws are tightened, and gives a firm grip at the front end of the jaws.

The surface of the lower clamps and the surface of the face plate are at right angles to the axis of the lathe; thus any hole or surface cut into a plate or bridge will be equally true.

A typical operation in which the face plate is of particular use is rebushing and jeweling the lower pallet hole jewel in which the setting is so badly damaged that any jewel or bushing set into it would not only be off center, but because of the short arbor heights, would cause the pallet to slant up, down, or sideways. This operation will now be described and serves to demonstrate how the face plate is used when encountering problems in an uprighting job.

Because of the requisite thickness of the pump-center, putting the pump-center through the gouged out hole of the lower plate to contact the jewel hole of the pallet bridge is difficult, and thus it is advisable to center this part by one of two other methods (and often both are combined to insure accuracy).

The first method employs the tailstock as a guide in centering the pallet bridge hole in the face plate. The second uses a pointed pegwood pivoted on a T-rest close to the plate with its point within the jewel hole. By observing the shimmying of the outer end of the pegwood as the lathe is in motion, the amount of error can easily be observed and corrected.

To use the first method, mount the face plate securely in the lathe and fit the adjusting rod of the pump-center until its point reaches out to a distance on the jaws where the pallet bridge will be when the movement is attached to the face plate. Advance the tailstock center-point towards this pump center-point and carefully observe the tailstock point from all angles to determine whether the points meet. If the tailstock point does not meet head on but is below that of the pump center-point, raise the tailstock point by putting a thin shim under the front parts of the tailstock, as shown in Figure 34. (A shim is a thin strip of metal, or other material, for fitting or bringing one part in line with another.)

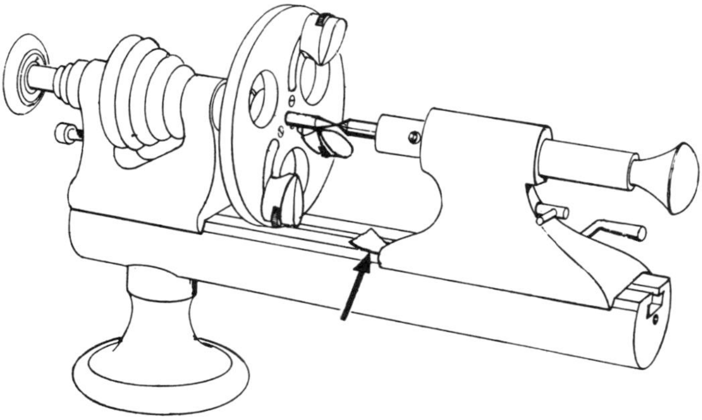

Fig. 34. Minor errors in off-center in the tailstock alignment can be corrected with the use of shims (see arrow).

Should the tailstock point be off center to one side, inserting a shim under the tailstock on that side will shift it towards the center. If the tailstock point is above the pump center-point, shims under the rear of the tailstock will tilt the front end down towards the center; should the tailstock point lie below and to one side, a shim at the front and on the off-side will rectify the error. This method can be used to correct an off-center tailstock center for minor faults so that its centerpoint will coincide exactly with that of the face plate in the headstock.

When the tailstock is checked and centered, *leave it secured in that position on the lathe bed,* although the spindle and center may be withdrawn temporarily.

FITTING MOVEMENT IN FACEPLATE

Fit the stripped movement with the attached pallet bridge facing the tailstock to the face plate, and clamp, not too tightly, with the three clamp jaws. Shift the plate with the attached bridge so that the

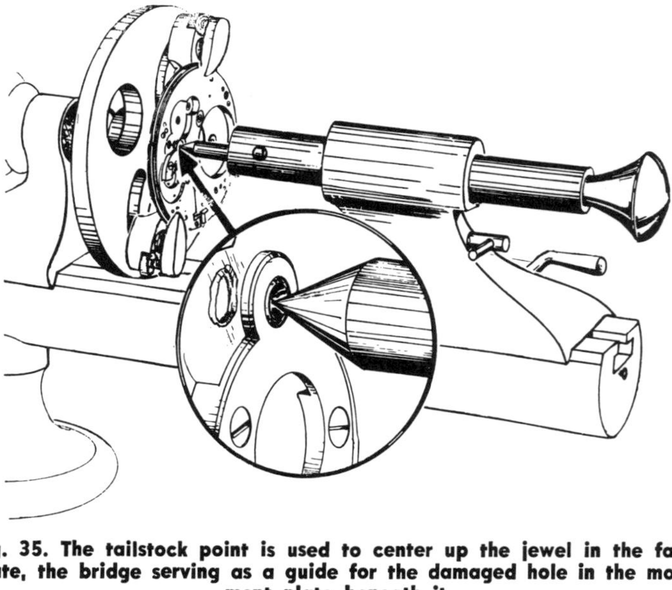

Fig. 35. The tailstock point is used to center up the jewel in the faceplate, the bridge serving as a guide for the damaged hole in the movement plate beneath it.

center of the pallet bridge jewel hole is directly over the point of the tailstock. Advance the tailstock to check on the alignment by loosening the clamp hold on the movement sufficiently to permit the plate to be shifted by the left hand. At the same time, the right hand can manipulate the tailstock spindle so that its front end point may be advanced or retracted.

With the eye loupe trained on the tailstock centerpoint, shift the plate until it can be "felt" and seen that the right hand has advanced the point into the jewel hole of the pallet bridge (Figure 35). The center point must enter the jewel hole gently, otherwise it may crack the jewel. After this is done withdraw the tailstock spindle and carefully tighten the clamps, a little at a time, and check to determine whether the plate has shifted while the clamps were being tightened. To make certain that the clamps grip the plate evenly and firmly, adjust the back nut ("M," Figure 33) by tightening the clamps moderately and turning the nut downward until it makes contact with the bottom of the recess in the lower clamp jaw. Tighten the lower clamp-post nut ("S"), which provides a firm grip on the plate.

Another method of centering the work in the face plate is by using the pointed pegwood mentioned previously. This method can also be used as a super-check on the accuracy of the first method (tailstock centerpoint). The pegwood is sharpened to a fine but strong point and this point is placed into the pallet bridge jewel hole. The T-rest is brought up close to the work and raised so that the pegwood resting on the T-rest is as close to the axis of the lathe as the eye can judge. If the jewel hole into which the pegwood is resting is off-center, the outer, rear end of the pegwood will shimmy perceptibly when the lathe is set into motion as shown in Figure 36. Minute errors in "off-center" will be multiplied or exaggerated due to the comparatively short distance from the jewel hole and the pegwood point in the fulcrum (T-rest) and the great length of pegwood beyond it. If the pegwood test reveals any degree of error, the plate may be shifted to satisfy a recheck. When the plate is perfectely centered, the outer end of the pegwood will remain motionless when the lathe is in motion. Some watchmakers use a horn-hammer to tap the not-too-tight plate in the direction of the indicated center.

If the pallet bridge hole or any of the bridge holes are gouged out (instead of those in the plate), the corrective process may be approached differently. In this case, the plate's perfect jewel (below that of the bridge) is used to obtain the uprighting center. The plate is placed on the faceplate *without* the damaged bridge and the jewel hole in the plate is centered. This can be done by

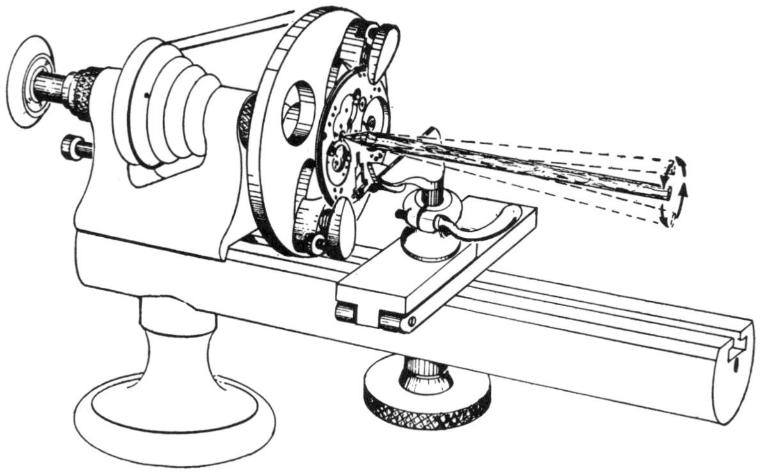

Fig. 36. Pegwood and T-rest, placed in position, will reveal any errors, in centering the movement plate on the faceplate, by shimmying of the end of the pegwood. If correctly centered, pegwood remains motionless.

the two methods just described or by a method to be described now which will also further illustrate the use of the face plate.

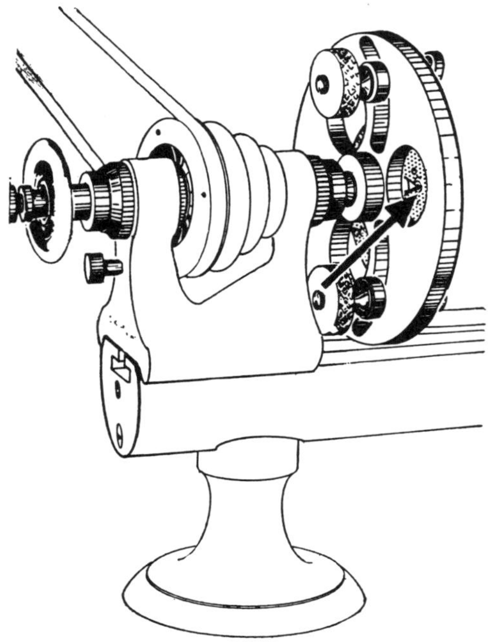

Fig. 37. The movement plate may be centered by observing the pump-center point from the rear of the faceplate through EACH of the three peep-holes.

First the bare movement is attached to the face plate, clamped near its edge by the three clamps, using the same precautions noted previously in regard to a tight grip. Also the position of the clamps on the plate must not interfere with the work to be done. Then the plate is shifted so that the center of the jewel hole is over the pump-center point of the face plate. This can be done by focusing the eye loupe on the pump-centerpoint through the peep-holes from the rear of the face plate as shown in Figure 37. The eye must be satisfied that the point is centered in the jewel by observing the point *through each of the three peep holes.* This is done by the left hand manipulating the pump-center-rod through the draw-in-spindle while the right hand carefully shifts the plate. The pump-center is delicately advanced until you know by the "feel" that the pump-center is in the jewel hole. Withdraw the pump-center into the face plate and tighten the clamps, and advance the pump-center again for re-checking through the three peep-holes. As a double check, apply the pegwood test, as shown in Figure 36.

Fig. 38. Boring tool is best held in slide-rest. Inset shows a boring tool made of thin drill-rod held in special holder.

When the accuracy of the center is assured and the plate is firmly secured, rebore the damaged hole. If this hole is in the plate, remove the bridge since it was used only to guide a center point, although the plate should remain in the faceplate. If the hole in the bridge needs correction, attach the bridge to the main plate *after* the center has been obtained without disturbing the center in the faceplate. Then rebore the damaged hole, removing only as much metal around the spoiled hole necessary to eliminate traces of previous damage inside the hole. (Thus the new hole is obtained by trimming off the scratches and tears inside the old hole).

The best method of boring a true hole is with a boring tool held in the tool post of a slide rest, or with a boring cutter held firmly in the hand. A boring tool is shown in the slide rest (Figure 38) with a special holder in the slide rest tool post. This slotted eccentric holder is designed to grasp round steel stock which can be easily shaped to cut objects held in the lathe (in this case a boring tool). A good hole may also be cut with a boring cutter set into a handle and held in the hand.

If the hand graver is employed, an L-rest is the best type of graver rest to use because the fingers can hold the graver, and approach the work more closely and with good control. The short section of the L-rest reaches into the plate between the clamp jaws. The pin on this section of the L-rest banks the graver or hand tool, and prevents it from sliding off the L-rest and hitting the jaws of

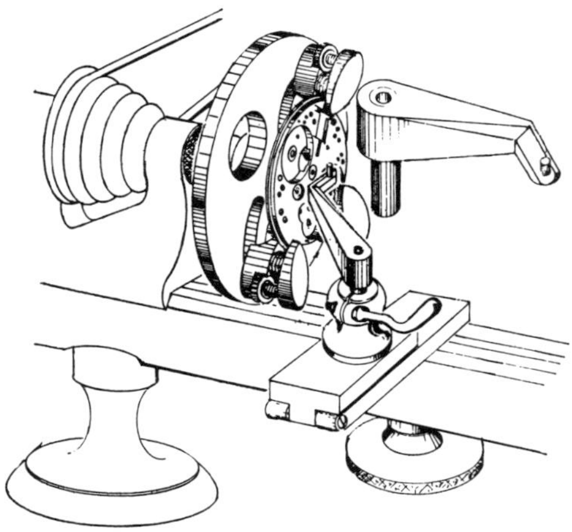

Fig. 39. If hand gravers must be used, an L-rest will permit close work between faceplate clamp jaws.

the face plate—and damaging the work or plate or causing personal injury (see Figure 39).

The rebored hole will be directly opposite that of the bridge or plate which was used to guide the center point. The center distances therefore are the same, allowing the pallet to remain level and at the correct distance from both the escape wheel and balance centers.

Fit the new hole with a jeweled bushing or with a new larger-diameter jewel, bezel-set into the plate—or press a friction jewel into the hole. Because correct horological practices require that the fine appearance of the watch be maintained, the replacement jewel should be the same type of jewel setting found in other parts of the watch as well as of the same type being replaced.

One other method of uprighting is to ream out the gouged out hole *before* it is placed in the faceplate, after which the hole is plugged with a solid bushing slightly tapered to match the taper of the broached hole. Press or tap the bushing friction tight into the hole (the bushing is made of the same color metal as in the plate). Set the jewel bezel style during the uprighting process. Instead of boring a hole, a seating, bezel, and countersink is cut for the jewel, pressed or burnished over, and the countersink stripped with a bright cutting graver.

By following the methods described, any jewel or bushing from any wheel or position may be centered, set, and uprighted.

Part III

FRICTION JEWELING

ONE of the greatest strides made in the field of horology during this 20th century is friction jeweling. This seemingly simple method, in which a jewel is merely pushed into a tight-fitting hole, had to wait until machines could be perfected that would produce perfectly round jewels of precise diameters. Machinery also had to be perfected for cutting holes in the movement of exact hole size and center distances. Today, the majority of watches of all qualities have jewels inserted by this method.

REPLACING A BEZEL-SET JEWEL WITH A FRICTION JEWEL

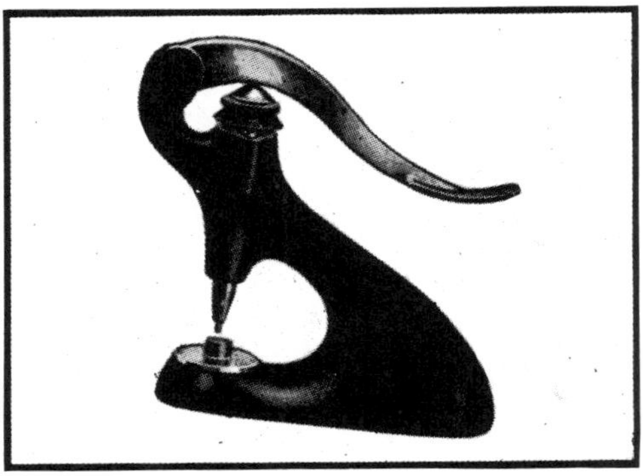

Fig. 40. A typical friction jeweling tool (Seitz) used by many watchmakers.

To replace a plate, balance, or cap jewel, use jewels and settings most closely in keeping with the method of setting and appearance

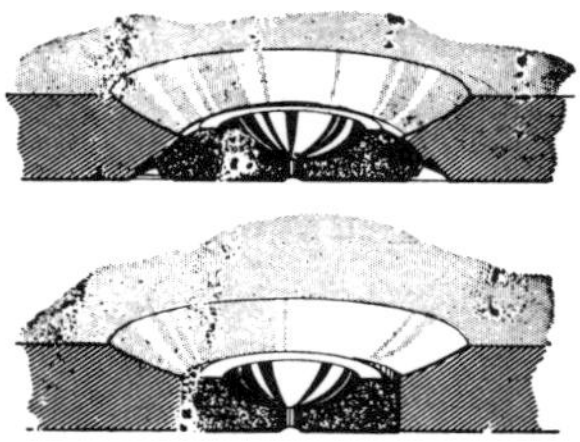

Fig. 41. Top: bezel-set jewel. Bottom: straight-walled friction-tight jewel.

of the other jewels in the watch. In older and cheaper watches, this is not always economically practical and thus a friction jewel may replace a bezel-type.

Fig. 42. Top: smallest friction jewel that can be fitted to reamed-out bezel setting (notice short wall area, indicated by arrows). Bottom: same setting reamed out to take friction jewel.

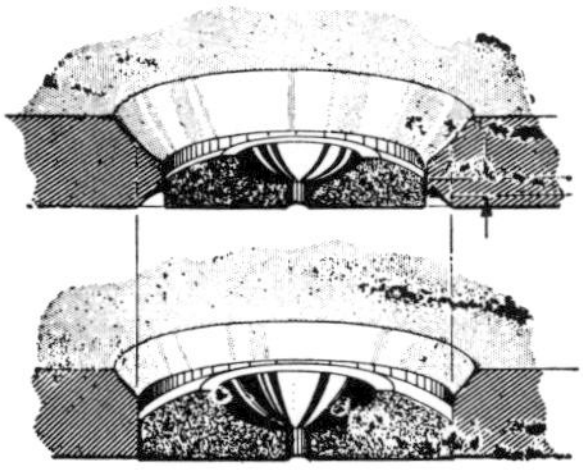

In a bezel-set jewel, the inside of the setting is not cylindrical and does not offer a sufficient surface on which the walls of the straight-side friction jewel may grip. It is easily seen from Figure

41 that to make the bezel setting accommodate a friction jewel, the bezel setting must be reamed out to a larger size and its hole made cylindrical. Figure 42 (top) shows the bezel setting already enlarged to take the nearest size friction jewel. Notice that the hole was enlarged just enough to take the friction jewel and that the outside bezel groove remains.

If this outside bezel groove does not leave sufficient tension around the jewel to grip it safely (often the case), ream the hole of the setting to a larger size (Figure 42, bottom). This operation

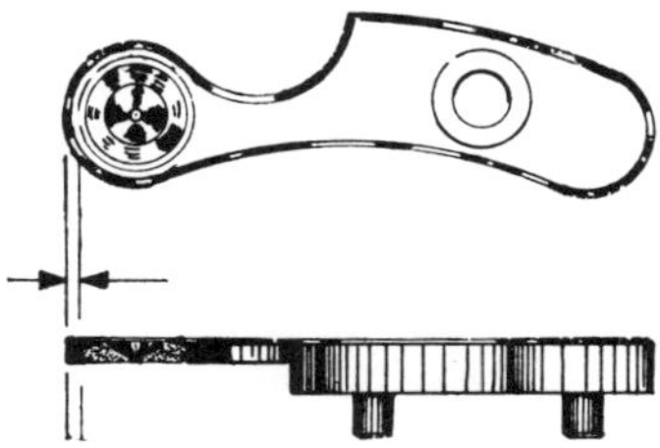

Fig. 43. When a pallet jewel or other type setting is close to an edge, replace with same type setting.

is generally permissible. However, when the bezel-set jewel and setting are very close to the edge of the bridge, as in some pallet bridges (Figure 43), and sometimes close to the lower center jewel setting, enlarging the hole may cause the thin wall to break. When this condition exists, the safest way of replacing a jewel without breaking the thin wall is by bezel-set methods described in the preceding section.

THE FRICTION JEWELING TOOL

If the bezel set hole is to be reamed to size, this must be done so that the wall of the hole is cylindrical and perfectly round. Also, this hole should then gauge 1/100 mm. *smaller* than the jewel to be fitted so that when the jewel is press-forced in, it will grip the jewel tightly. Ordinary broaches certainly cannot fill these requirements. Furthermore, it is obvious that even an accurately designed reamer will have to be guided in a specially designed tool so that the hole is plumb and on the same axis as the pivot which the jewel will serve as a bearing. There are many different makes of friction jeweling tools but all of these work on the same principle. A typical type of friction jeweling tool is shown in Figure 44. This is shown in a part cutaway and exploded view so that it can be better understood.

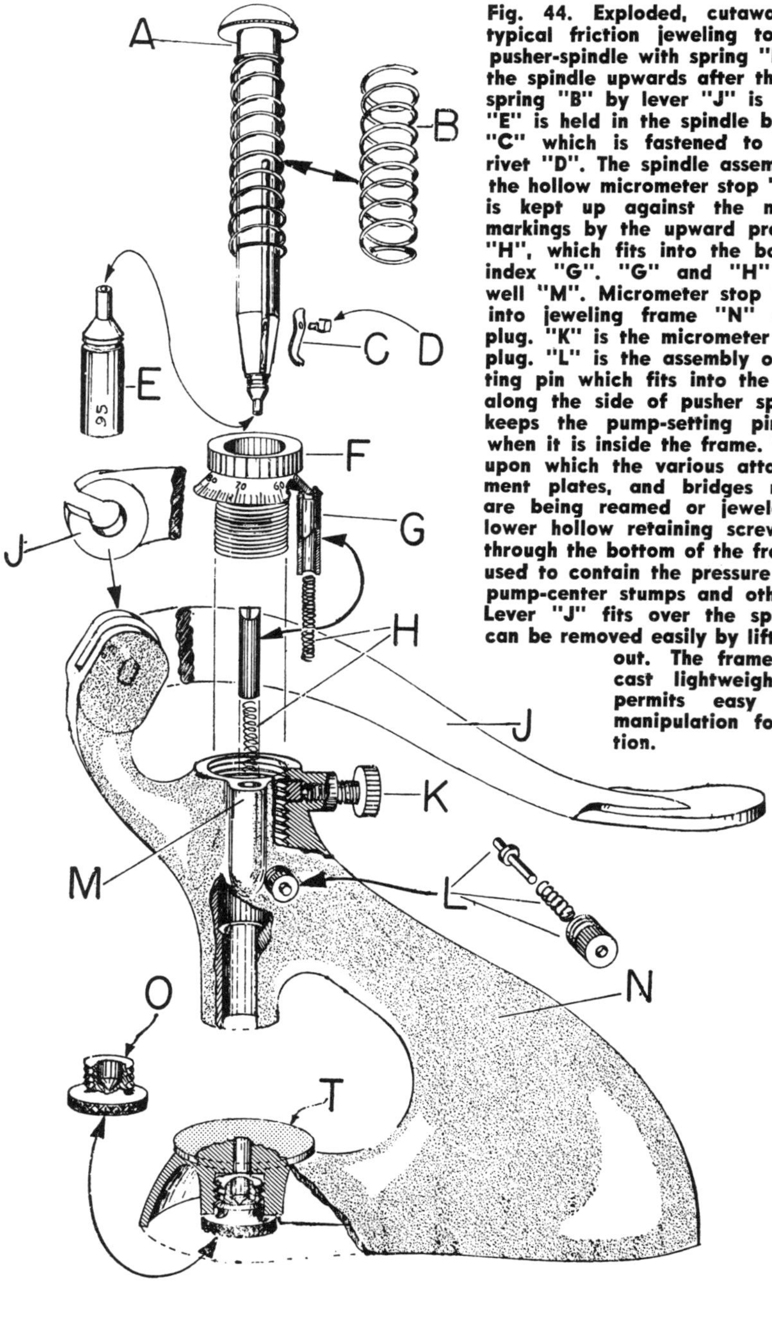

Fig. 44. Exploded, cutaway detail of a typical friction jeweling tool. "A" is the pusher-spindle with spring "B" which pushes the spindle upwards after the pressure upon spring "B" by lever "J" is relieved. Pusher "E" is held in the spindle by tension spring "C" which is fastened to the spindle by rivet "D". The spindle assembly fits through the hollow micrometer stop "F". Index "G" is kept up against the micrometer stop markings by the upward pressure of spring "H", which fits into the bottom of hollow index "G". "G" and "H" fit into index well "M". Micrometer stop "F" is threaded into jeweling frame "N" on the friction plug. "K" is the micrometer lock screw and plug. "L" is the assembly of the pump-setting pin which fits into the groove running along the side of pusher spindle "A", and keeps the pump-setting pin from turning when it is inside the frame. "T" is the table upon which the various attachments, movement plates, and bridges rest while they are being reamed or jeweled. "O" is the lower hollow retaining screw which fits up through the bottom of the frame base and is used to contain the pressure springs used in pump-center stumps and other attachments. Lever "J" fits over the spindle head and can be removed easily by lifting upward and out. The frame of the tool is cast lightweight metal which permits easy handling and manipulation for close inspection.

To fit a new jewel, use a jeweling tool, a flat stump upon which the bridge or plate will rest, the pushers which will force in the jewel, and the reamers which may have to be used to prepare the hole for the new jewel.

If the old jewel in the movement is not badly damaged, use it as an aid in adjusting the stop micrometer for proper endshake. Select the largest flat stump upon which the plate, with its inside surface up, may rest flat, and insert a flat pusher (slightly smaller in diameter than the jewel it will contact) into the front end of the spindle. Insert the spindle and pusher into the frame of the jeweling tool and rest the pusher on the damaged jewel. Turn the micrometer stop so that it just makes contact with the underside of the spindle head. This sets the distance to which the pusher later may be pressed down.

Remove the old jewel by methods described earlier. If the old jewel was bezel-set and is to be replaced with a friction jewel, ream out the hole.

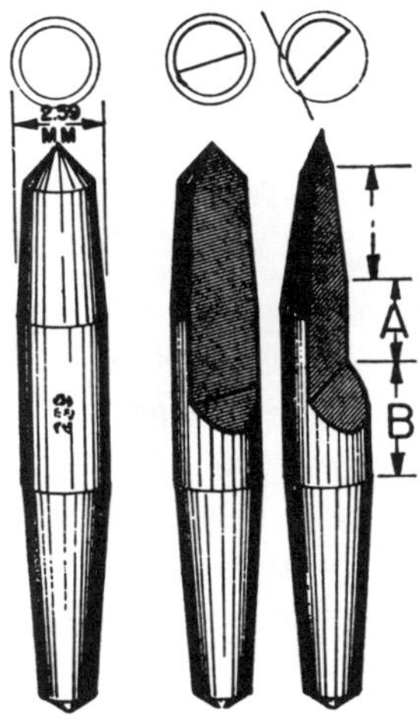

Fig. 45. Step-by-step construction of special reamer (marked in millimeters) is shown in the illustration at the left. The section above "A" is the tapered broaching portion. "A" is the straight reaming area. "B" is the finishing part (to bring the hole to correct diameter).

Most jeweling tools come with a set of reamers for preparing a smooth cylindrical hole. The reamer (Figure 45) is a slightly tapered, polished steel rod which has had half of its side ground off above the taper so that part of the cutting edge is cylindrical. The lower cylindrical cutting edge provides the finishing touch to the hole.

If the hole must be reamed to provide a walled surface for the friction jewel, the reaming operation is done with the plate or bridge resting on the face of the tool, inside surface upwards. The reamer is held in the reamer handle shown in Figure 46. This handle is of the same diameter as the pusher spindle and fits into the jeweling frame. When this reamer or "smoothing broach," as it is sometimes called, is lowered into the setting, the handle is twisted by the fingers with a slight downward pressure, the fingers of the other hand holding the plate or bridge rather loosely so that the bridge or plate may follow the reamer as it is being turned within the hole. This is done so that the hole may be influenced only by the cutting of the reamer and thus result in a true and round hole. In this way the reamer will cut away the walls of the old setting as shown in Figure 47. The reamer should always be twisted so that it cuts gradually and it should also be turned and

Fig. 46. Back-tapered reamer fits into broaching or reaming handle.

twisted until the cylindrical portion of the reamer is within the hole it had just cut. If the hole must be enlarged considerably, this will have to be done in gradual steps, using first the smallest reamer to fit the hole, then the very next larger size and so on until the desired size is reached. To "short-cut" the gradual opening of the hole by forcing the cutting with a larger reamer invites the cutting of a hole which is egg-shaped, creating off-the-center distances with the added possibilities of breaking the reamer.

As mentioned before, the holes must be slightly (.01 mm.) smaller than the jewel it will hold. This is necessary so when the jewel with its chamfered edge is press-forced into the hole, this expanded hole will grip the cylindrical side of the jewel tightly.

Each reamer, although its front is slightly tapered, has a precision cylindrical finishing section and is stamped with its gauge. Thus the reamer marked 119 (1.19 mm.) cuts a cylindrical hole of that size and the jewel to fit must measure 1.20 mm. in diameter. After a jewel hole has been reamed, all burrs which were raised should be removed without removing any metal below the surface of the plate.

FRICTION JEWELS

The bezel-set jewel was designed to be held by a rim of metal pressed over its edge and a bevel for this purpose was provided.

Fig. 47. Watchmaker carefully reams hole for jeweling, handling reamer and holding plate (left). Enlarged view of reamer (right) shows that reamer cuts through the hole so that the reamer's cylindrical smoothing section is within the hole.

In a friction jewel, on the other hand, the gripping surface is the cylindrical wall of the hole. Keep this in mind when setting a friction jewel.

Figure 48 shows three types of friction jewels: olive-holed domed-balance jewel; straight-holed plate jewel; cap jewel. (The parts of the jewels are labeled for future reference.) Notice that balance and plate jewels have the chamfered or beveled edges on the same side as their oil-sinks. When inserting jewels such as these two, press with the oil sink down (to start the jewel into the hole). Use a pusher only slightly smaller than the diameter of the jewel, otherwise one or both edges of the pusher may contact the edge of the setting, preventing the jewel from being pressed to the correct depth; an over-sized pusher may also damage the setting by scoring its sides.

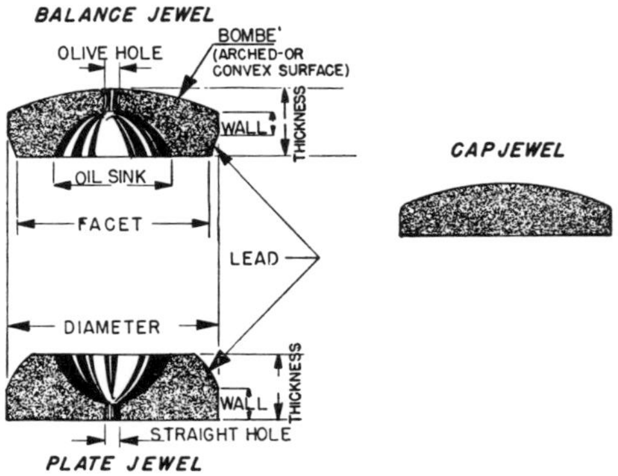

Fig. 48. Types of friction jewels—balance, plate, and cap—and their component parts.

Figure 49 shows a plate jewel being pressed into a train bridge. Notice that the jewel is inserted from the bottom of the bridge (which is facing upward) so that the chamfered edge of the jewel helps the jewel into the setting. The oil-sink faces the top of the bridge.

When setting a flat-faced jewel flush with the plate, the watchmaker may use a pusher considerably larger in diameter than the jewel and the setting (Figure 50) and thus obtain a firm, flat seating for the jewel (flush with the plate).

When pressing in a jewel, the micrometer stop governs the distance the jewel is to be pressed into the plate. However, press the lever firmly so that the jewel is inserted the full depth of the micrometer allowance. Adjust endshake by twisting the micrometer feed screw.

BALANCE JEWELS

When replacing a bezel-set jewel with a friction jewel, use friction jewels and settings that are as close as possible to the setting and appearance of other jewels in the watch. Among the three types of friction jewels (described previously) which may replace the bezel-set jewel is the domed balance jewel.

Most balance jewels made today are "bombe"—their top surface is domed so that the capillarity of the oil (in its contact with the

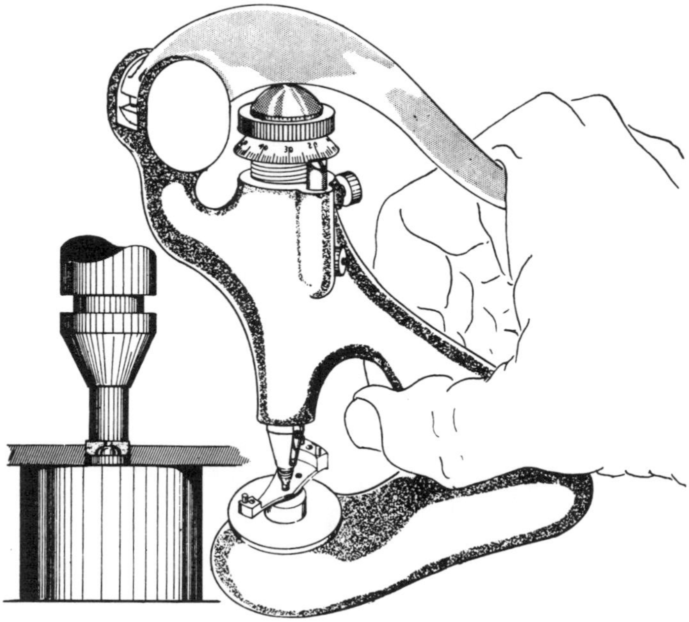

Fig. 49. How a plate jewel is pressed into train bridge. Note micrometer stop against pusher-spindle-head. Endshake is adjusted with this stop.

domed jewel and the flat cap jewel) will cling to the hole where it is needed. When this type of jewel is inserted, use a pusher which makes proper contact with the domed surface. Figure 51 shows a "concave pusher" (one of a set which is supplied with most friction

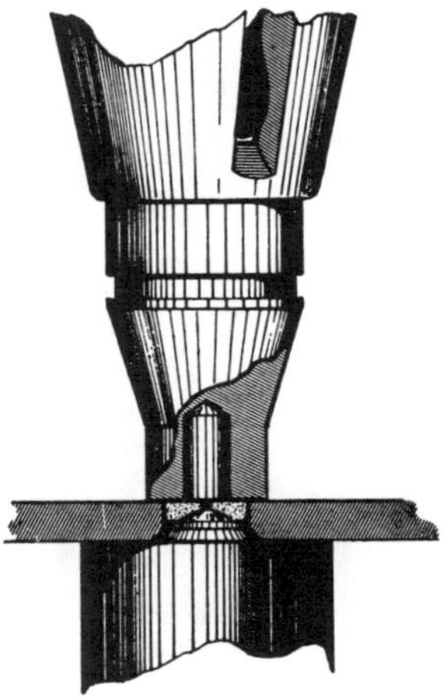

Fig. 50. When a flat-top jewel must be pressed in flush with plate's surface, use a large diameter (flat) pusher.

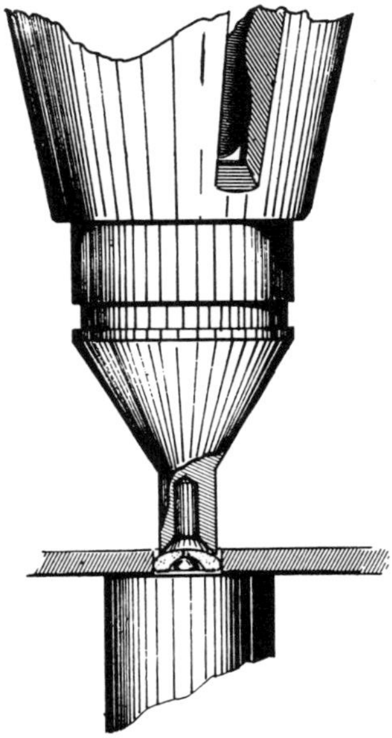

Fig. 51. Pusher presses in convex balance jewel.

jeweling tools) inserting a convex or ''bombe'' balance jewel. This type of pusher distributes its load of approximately 15 pounds pressure (amount required to press such a jewel into a hole .01 mm. smaller than the jewel) evenly on the jewel without danger of breaking the jewel.

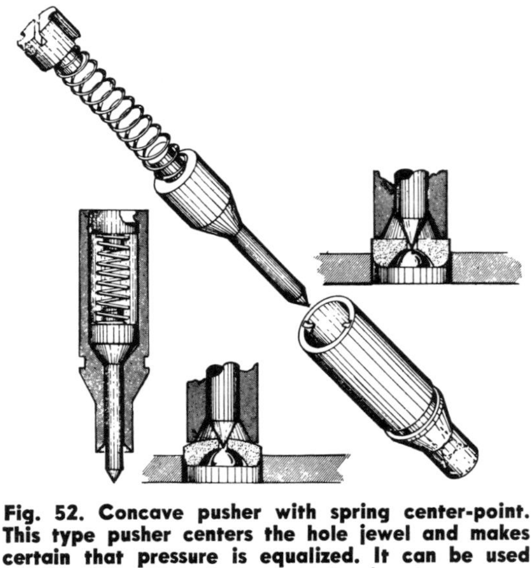

Fig. 52. Concave pusher with spring center-point. This type pusher centers the hole jewel and makes certain that pressure is equalized. It can be used with both convex and flat jewels.

Figure 52 shows a cross-section of a pusher with a spring centerpoint. This type is very desirable because the jewel is automatically kept in the center of the pusher, permitting more precise placing of the jewel in its hole (without the probability of the jewel shifting to one side and not lying flat after it is set). A pusher with a spring centerpoint may be used to insert both convex and flat hole jewels. Through constant use, this pusher with its bell-shaped surface becomes distorted and can cause jewel breakage unless it is disassembled and repaired. For this reason, these pushers are constructed so they easily can be taken apart; make a one-fourth turn of the slotted top plug (releasing the spring), and remove the centerpoint through the rear of the pusher. Sharpen the centerpoint on a lathe and grind the pusher to the required shape.

ADJUSTING ENDSHAKE IN A FRICTION JEWEL

If a wheel or pinion has too much endshake, the wheel or pinion can easily be adjusted to the desired tolerances with the friction jeweling tool: (1) Remove the bridge or plate from the movement and place it on a flat stump (anvil); (2) fit a pusher of slightly smaller diameter than the jewel in the pusher spindle; (3) lower the pressing lever gently until the pusher makes contact with the surface of the jewel; (4) raise the micrometer stop until it touches the

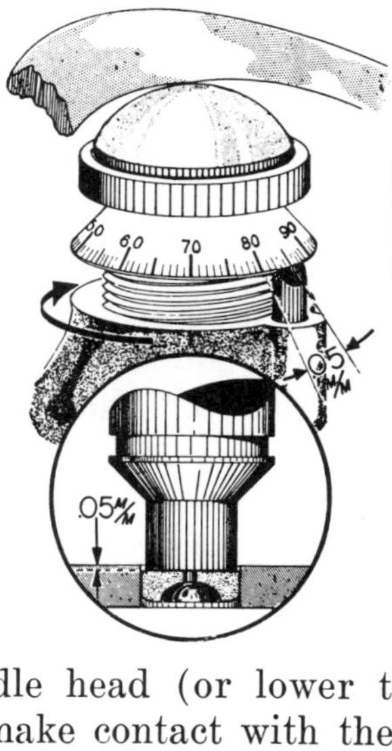

Fig. 53. Each mark on the micrometer allows a vertical movement of .01 mm. to the pusher. Arrow indicates direction of turning.

underside of the spindle head (or lower the micrometer feed to permit the pusher to make contact with the jewel). For example, to reduce the endshake .05 mm., turn down (clockwise) the micrometer feed 5 notches (Figure 53). This permits the pusher to go down that much lower, forcing the jewel deeper into the setting.

INCREASING ENDSHAKE

The operation of increasing endshakes or removing a friction jewel are much the same. When increasing endshake, push the jewel outward. To accomplish this, select a hollow stump (anvil) with a hole *slightly larger* than the diameter of the jewel and place the bridge or plate on this stump so that the oil sink of the jewel faces the pusher (upward).

Fig. 54. Select a pusher that is of the same diameter as the facet of jewel being pushed out.

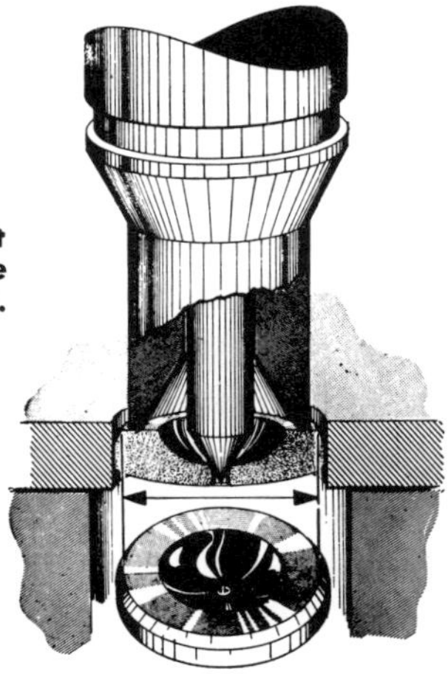

In selecting a pusher to force the jewel *outward,* choose one with a diameter as wide as the facet of the jewel (Figure 54). A larger pusher may break the jewel and ruin the setting; a smaller pusher enters the oil sink and may break the jewel.

Lower the pusher gently to contact the jewel and turn the micrometer stop until it contacts the pusher spindle head. Turn down the micrometer stop as required to increase the endshake, and press the push-lever to push the jewel outward. This operation actually is the same as decreasing the endshake except that the bridge or plate is reversed on the hollow stump.

In case the jewel must be removed, adjust the micrometer stop to permit only as much thrust as necessary, otherwise the pusher may go too deep and its tapered back section will plunge into the hole, ruining the plate or bridge.

REPLACING A LOOSE FRICTION JEWEL

Sometimes a friction jewel becomes loose, or, when a new friction jewel is fitted to a hole as a replacement, the hole often fails to return to its original size (.01 mm smaller than the jewel). In such

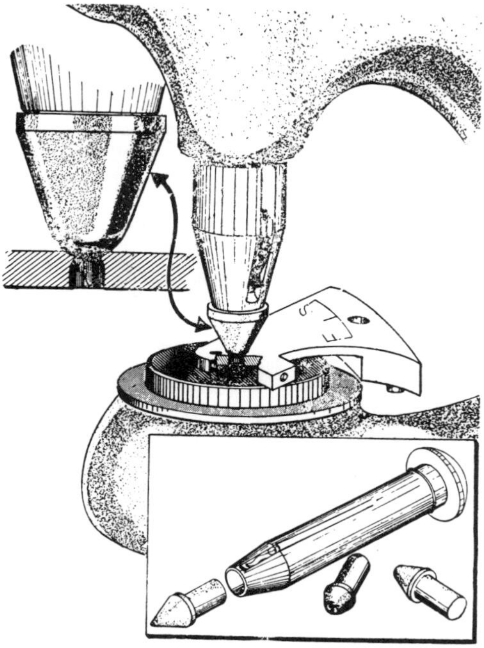

Fig. 55. If a jewel is slightly loose in its hole, reduce the size of the hole with solid-domed pushers, shown above. Inset—enlarged view of pushers.

a case, reduce the size of the hole, avoiding reaming the hole to a larger size and fitting a new (larger) jewel.

For this hole-reducing purpose, some jeweling sets are supplied with hardened, solid-rounded top pushers (in Fig. 55 one of these is shown compressing the hole of a bridge). In this operation, avoid too much pressure when closing the hole because the walled surface might be made too thin. Also, after reducing the hole, ream it with the proper reamer to restore it to the exact hole size and insuring that the sides are cylindrical instead of tapered (which might cause the jewel to crack when it is forced in). Re-insert the jewel in the same way as a replacement, using the same precautions and operations as in putting in a new jewel.

FITTING FRICTION JEWELS IN SMALL BRIDGES AND CAPS

Should a pallet bridge or a cap jewel require reaming or a new jewel, holding it in the fingers upon the face of the jeweling tool would be awkward and difficult. To facilitate these jobs, most sets come with a special lightweight plate which is equipped with a series of clamps that can be used to hold such smaller objects while

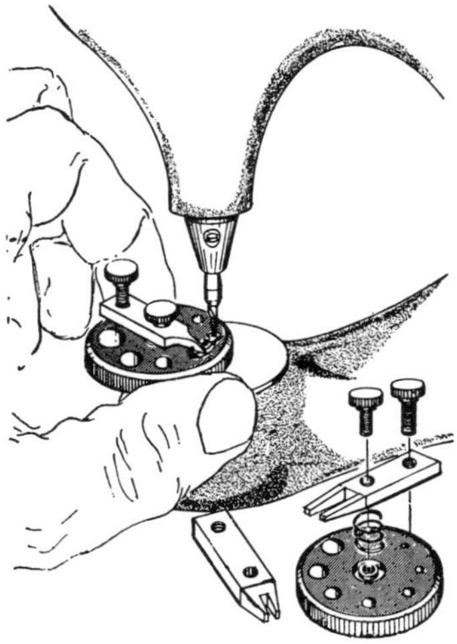

Fig. 56. Light block and clamps to hold pieces to be reamed.

these are being reamed or jeweled. In Fig. 56 is shown such a plate and clamp holding a pallet bridge. The reason for the plate being of lightweight material is that it will follow the reamer should this have a slight bend, thus avoiding the possibility of cutting a hole which is not perfectly round. Clamp the part tightly, so that these small pieces will be perfectly flat for the pressing-in of the jewel.

SETTING FRICTION JEWELS IN BUSHINGS

When a brass, jeweled bushing needs a new jewel, it may be pressed into the bushing while it is held in the plate or bridge, providing the bushing is tight in the plate. Should it be loose and require reaming or a new jewel, a set of jewel chucks with recessed ends is available. For the friction jeweling tool, these are especially made very short and unthreaded as shown in Fig. 57. These fit into a tapered hole in a round flat plate which rests on the face of the friction jeweling tool. The tightening action of the chuck is supplied by a clamp whose front end is chamfer-holed out to fit over the convex top of the chuck. Tightening the clamp with the thumb screws causes the clamp to press down on the chuck which in turn makes the chuck close its jaws around the bushing. The reaming and jeweling is done through the hole in the clamp tip.

The jewel chucks should be cautiously tightened less they crush the bushing.

When reaming out the hole in the setting as in other pieces, the plate holding the chuck bushing is grasped lightly so that the plate may follow the slight divergencies of the turning reamer. If the plate were fixed to the face of the jeweling tool frame, the holder

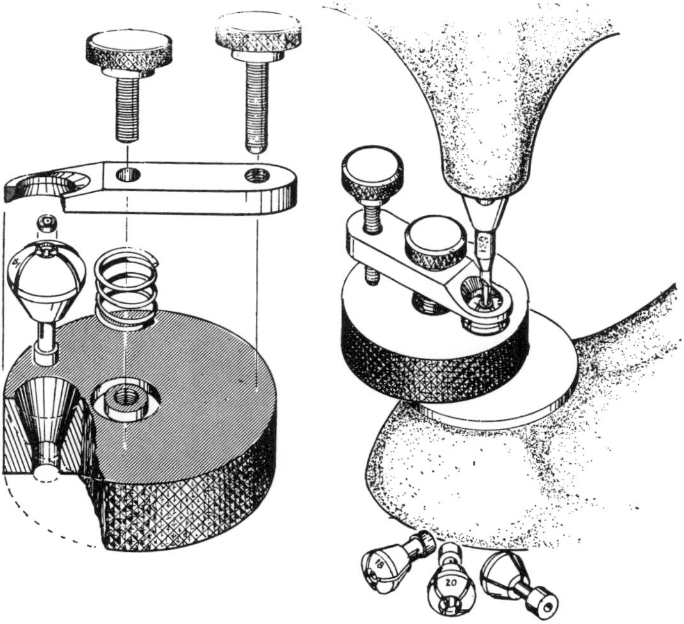

Fig. 57. Block and jewel chuck attachment with some different size jewel chucks which are used to hold bushings while they are being reamed. Set comes with 12 different chucks.

would not be able to shift with the reamer and the hole might be reamed out-of-round.

In this manner, bushings formerly set with bezel-type jewels may be reused with friction jewel replacements, reaming out successive diameters gradually, until the desired hole size is obtained.

UPRIGHTING WITH THE FRICTION JEWELING TOOL

The best method of uprighting an off-center hole is to place the movement in the face plate of the lathe and perform the job as described earlier. However, once the center and hole have been obtained, it may be reamed and jeweled with the friction jeweling tool. Some friction jeweling devices come equipped with a face plate, in which case the movement is mounted in it and the face plate placed into position in the friction jeweling tool. Then a centering tool or centerpoint is placed into the pusher spindle. If a lower (plate) jewel must be uprighted, the (upper) bridge is attached and the movement shifted on the face plate so that the good jewel is directly below the centering point. The centering point is then allowed to carefully enter the hole of the jewel which will be used as a guide. The face plate clamps are tightened, center re-checked and then the bridge is removed without disturbing the movement's position in the face plate. The lower hole is reamed and then jeweled, the jeweling taking place after the movement has been removed from the face plate.

If the hole in the bridge must be uprighted, then the hole in the lower plate is aligned with the centerpoint, the bridge later attached, reamed (and jeweled separately). Fig. 58 shows the type of face plate used with the friction jeweling tool.

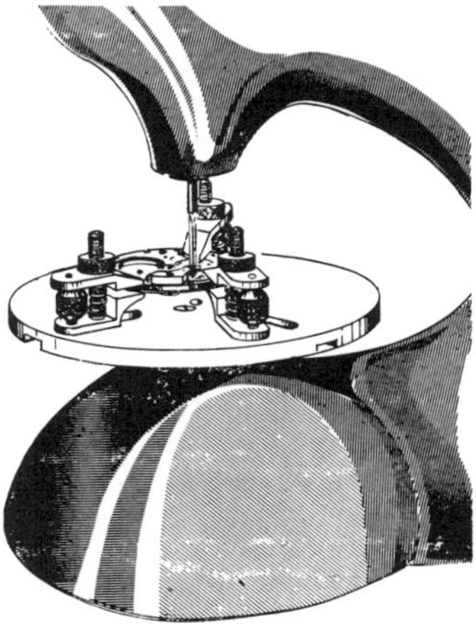

Fig. 58. Some jeweling outfits come equipped with a face plate, such as shown at left holding a plate and bridge for uprighting.

Some friction jeweling tools come supplied with pump centerpoints which fit up through the base of the tool and are supposed to aid in guiding the plate for the uprighting operation. This is shown in Fig. 59. However, in uprighting by this method, the

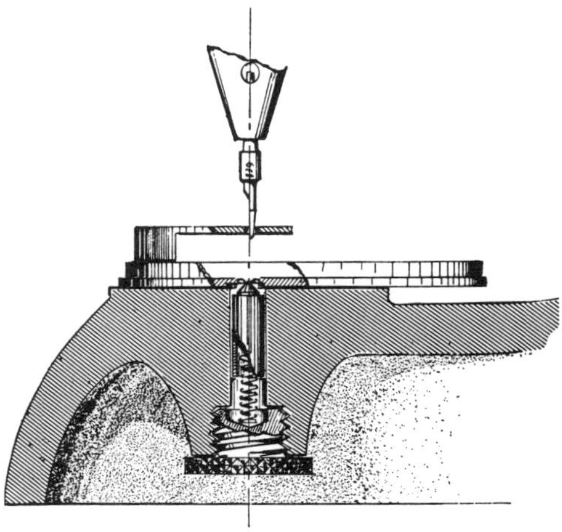

Fig. 59. Pump centerpoint, which fits in base of tool, used in uprighting. Using the pump centerpoint is not as reliable as using a face plate for uprighting.

movement will have to be held rigidly by the fingers while the reamer from above corrects the uprighting. This method is hardly as efficient as that which employs the use of a face plate rigidly fixed and centered.

Fig. 60 shows a base with assorted spring-loaded centerpoints. It is composed of the base and hollow stem "B"; the centerpoint guide bushing "A"; the centerpoint "D"; the spring "E"; and the spring plug "F." The centerpoints come in assorted thicknesses, each with its own matched bushing. To assemble the attach-

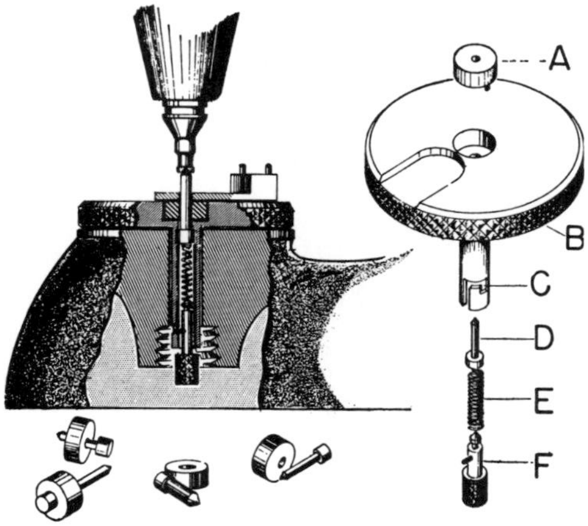

Fig. 60. Spring loaded base designed for the insertion of jewels on mass production-repetitive basis. "A" is the precise base which fits into the counterbore in the base "B". The guide "D" fits up through "B", through "A." The spring "E" supplies the tension on the guide "D". "F" is the plug which holds the spring. The pin in "F" locks the plug in the groove "C" of the post of the base.

ment for use, the plug "F" is pushed upwards and twisted to release its pin from the slot "C," then the centerpoint "D" is inserted through the tube of the base "B." The spring "E" and the plug "F" re-inserted and locked in the notch "C." The matched bushing "A" is then placed over the protruding centerpoint and rested into the counterbore in the face of the base.

The figure at the left in Fig. 60 shows this tool in use. The bridge to be jeweled is placed over the centerpoint which fits the jewel bushing hole closely. The jewel is rested on the tip of the centerpoint. The pusher is shown contacting the top of the jewel and thus the jewel is guided into the bushing hole. This attachment is particularly useful where mass production-repetitive jeweling is desired on a particular bridge or plate. Although it is intended for manufacturing and mass assembly, its design lends itself to repair methods as well.

The jeweling tool is also adaptable as a staking tool for putting on hands or bushings. Some jeweling tools come with attachments for the staking of staffs, but staking is accomplished best in a staking tool whose base is specifically constructed to take the heavy treatment associated with staking on staffs, pinions and other heavy work.

The whole system of friction jeweling is based on a jewel that is perfectly round and of exact diameter, fitting precisely with pressure into a hole .01 mm (1/100 mm) smaller than itself. A hole 2/100 mm. smaller will cause the jewel to break. In a hole of the same size as the jewel, the jewel will be loose. Before using a friction jeweling tool on a "live" job, it would be advisable to practice on older movements, replacing the jewels and practicing on adjusting endshakes.

BOOK III

CHAPTER II

REPLACING A BARREL HOOK

WITH the availability of replacement materials for current popular watches, the younger watchmaker may become neglectful of the skills needed to make repairs to watches for which materials and replacement parts are unavailable.

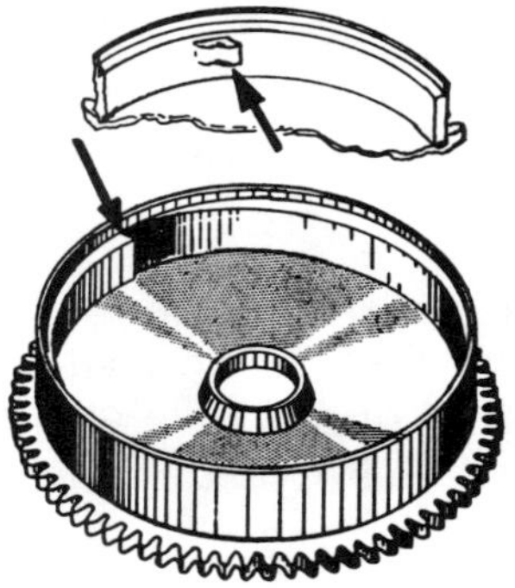

Fig. 1. (Top) A hook is used to brace either a tongue-end or a hole-end mainspring; (bottom) a tongue-end brace is milled out of the barrel wall.

Fig. 1 shows two types of anchorages for the mainspring. The upper figure shows a hook type and is used to fasten either a hole-end or tongue-end mainspring. Such a hook is either milled out of the barrel blank at the time of manufacture or it may be a separate piece, threaded or riveted in place. The arrow in the lower illustration points out the most common type of mainspring anchorage for a tongue-end (Swiss) mainspring. This type is milled out of the thickened barrel wall and is undercut around its lower edge.

Occasionally, these will either wear off and in the case of the hook, it may tear off due to a faulty mainspring endpiece or frequent mainspring slippage. Attempts to repair these will often result in damage to the barrel itself. Therefore, it is better to provide a new hook.

If the hook to be replaced is for a hole-end mainspring, the replacement must be fitted so that the new hook will align with the mainspring's hole. Should the anchorage for a tongue-end spring become worn, a new anchorage may be provided by fitting to it a hook as shown in Fig. 1.

The new hook should be fashioned from a brass taper pin, threaded and fitted through a hole drilled and tapped through the barrel wall. Since the barrel wall is generally very thin, there is not too much support for the threaded pin. Therefore, the hole should be drilled diagonally through this wall. Fig. 2 shows two holes, one drilled radially and the other diagonally. Notice that the diagonally drilled hole will provide a greater threading surface than the radial hole. Also because of its angle of inclination, it offers a stronger brace for the mainspring as well.

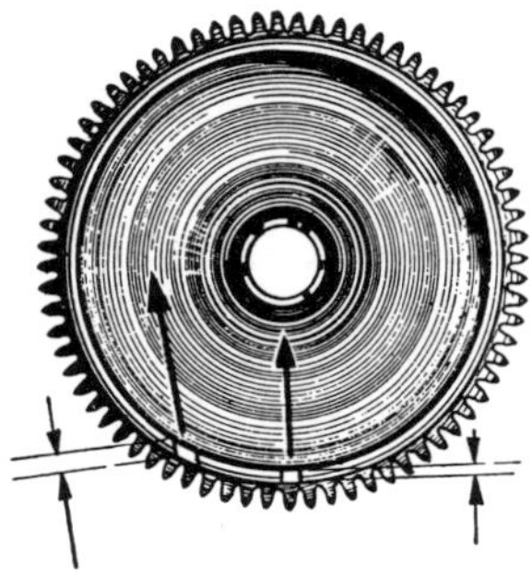

Fig. 2. More threading surface is obtained by drilling a hole diagonally through barrel wall.

To drill the hole, the blade of a small jewel-screwdriver is removed and replaced with a pivot drill of the desired size as shown in Fig. 3. With this tool, the beginning of the hole on the curved wall of the barrel can be easily spotted and drilled. The spot where the hole is

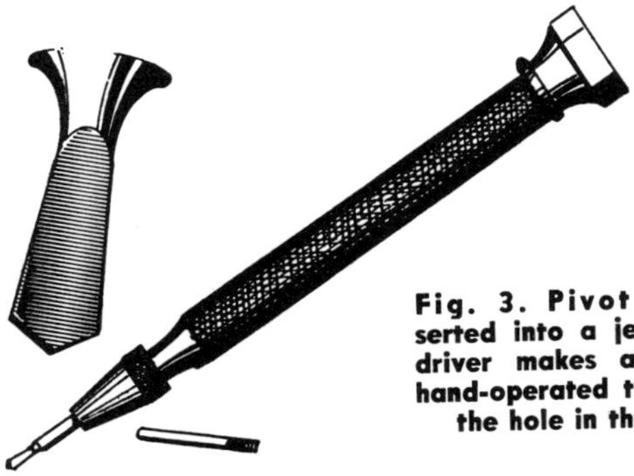

Fig. 3. Pivot drill inserted into a jewel screwdriver makes an efficient hand-operated tool to drill the hole in the barrel.

drilled on the barrel wall is important. This hole should be made at a point midway from the bottom of the inside of the barrel and the cover-ledge of the wall. To judge this distance from the outside of the barrel is difficult as the barrel teeth require considerable height and can easily deceive the eye. To start the diagonal hole, the drill must first be pointed towards the center of the barrel and turned until the width of the countersink made by the drill-point is equal to the diameter of the drill as shown in Fig. 4. Then, the

Fig. 4. Before a diagonal hole can be drilled into the barrel, a countersink in a radial direction must be made.

direction of the drill is shifted to form the diagonal hole. If the spring develops counterclockwise (from the endpiece towards the center) as in most barrels, the hole should be drilled in a clockwise direction, shown in Fig. 5.

Some watchmakers thread (tap) the hole in the barrel after it is drilled but this is not always necessary as the threaded taper pin itself will do that as it is twisted through the hole. Furthermore, since the hole size increases as the threaded taper is forced in, the

Fig. 5. After the countersink is cut, the drill's direction is shifted and the hole drilled through.

initial threading of such a hole is useless. Also, as the threaded taper forces itself in more deeply, the tighter it will become in the hole, creating the necessary threads.

The hook is made from a hardened brass-pin which is filed to a very gradual taper, after which it is threaded along a section of its length which will be fitted to the barrel. The thick end of the pin must be sufficiently thick to permit filing and shaping for the hook. Threading a taper-pin is easily accomplished by placing the taper-pin through a screwplate hole and threading it as far as it will go as shown in Fig. 6. It is then withdrawn and threaded through the next larger hole, repeating the first threading operation and so on until the threading operation is completed. Most screwplates have graduated holes in half sizes, making the taper-thread a very gradual one. The finished taper pin, threaded, should look like that

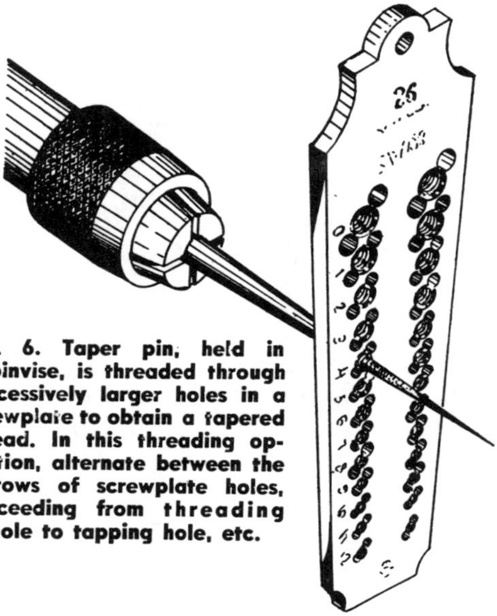

Fig. 6. Taper pin, held in a pinvise, is threaded through successively larger holes in a screwplate to obtain a tapered thread. In this threading operation, alternate between the 2 rows of screwplate holes, proceeding from threading hole to tapping hole, etc.

in Fig. 7. The front tip of this pin is left unthreaded so that it can be grasped in a pinvise.

Some watchmakers prefer to thread the taper pin into the barrel hole from the outside, claiming that the pull of the mainspring will tighten the taper-threaded hook instead of the possibility of loosen-

Fig. 7. Thread the taper pin only slightly more than necessary, to appear as shown when finished.

ing it. However, the better watchmaker reasons that the greater danger to the hook becoming loose is when the mainspring breaks.

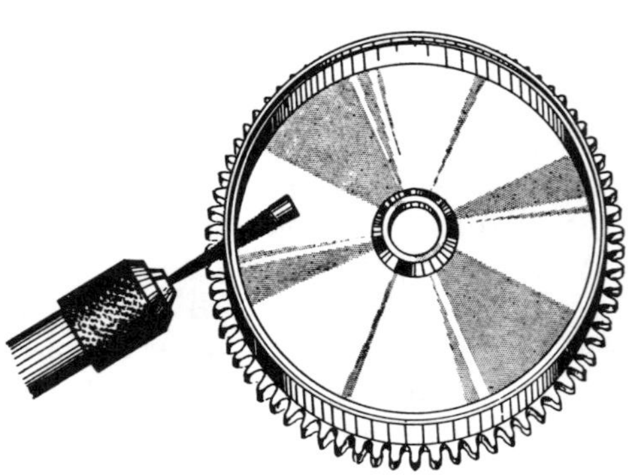

Fig. 8. Insert threaded taper through the inside of the barrel. Grasp the thin end in a pinvise and twist outward until tight.

In such a case, if the taper is thicker on the inside, a violent surge of (released) power occurs during breakage and would not eject the hook. The pull of the mainspring compared to this is rather weak and the greatest strain takes place only when the spring is fully wound. With the hook-plug facing diagonally, this strain is greatly relieved. For these reasons, the taper is inserted from the inside of the barrel outward.

After the threaded taper is inserted and its thin end is pointing outward, this section is grasped in a pinvise as shown in Fig. 8. Using this taper as a tap, twist it in and out until it is felt by the fingers holding the pinvise that the taper has been threaded as far as it can go without twisting and breaking off. It should be noted where the thick end of the taper should be cut off to serve as the hook. This is done by running the tweezer across the excess threads, scratching them. This also serves in locating which is the top of the taper so that when the hook is fashioned, it will point in the correct direction when re-inserted. This is shown in Fig. 9.

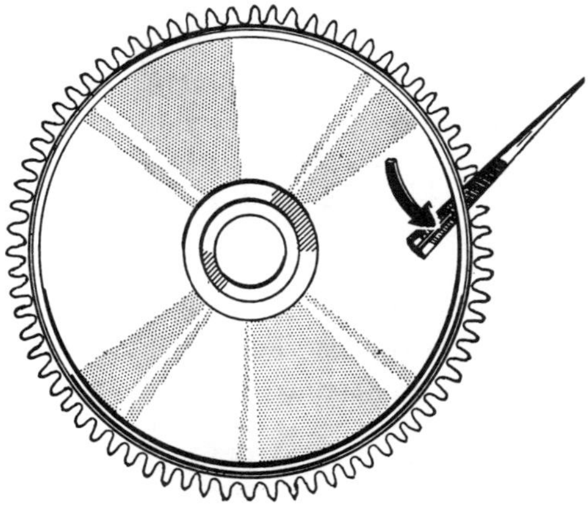

Fig. 9. Scratching the threaded portion inside the barrel indicates the point on the taper where the hook must be shaped.

The hook will then assume a position shown by the dotted line of Fig. 10. If the hook is to be used for a hole-end spring, it will have to be shaped to fit the hole and should look like that in Fig. 11. The amount that the hook protrudes from the barrel wall should not exceed twice the thickness of one coil of the mainspring. Should the hook protrude too far, it may kink the second coil when the spring is unwound. However, it should also be long enough to provide a grip for the endpiece.

If the hook is to provide a brace for a tongue-end spring, the hook need have no special shape other than being slightly undercut to provide a positive anchorage for the tongue-end brace.

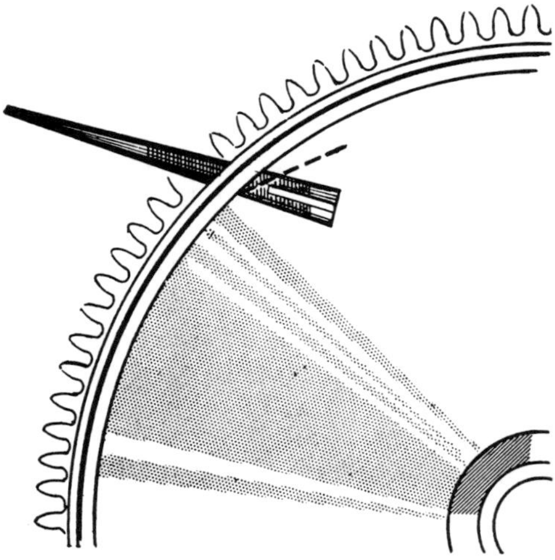

Fig. 10. To obtain a secure brace for the mainspring, file the hook (in direction shown by the dotted line).

When the taper is re-inserted after shaping its hook, it must be turned in tight with the hook facing in the correct direction needed to catch the spring. The taper outside of the barrel is then cut off close to the outside of the barrel wall as shown in Fig. 12. To trim this further, a file with a safe edge is used so as not to harm the barrel teeth.

This should not be filed flush with the barrel wall, but rather, a little bit should be allowed to remain. This excess is then burnished over with a fine file whose cutting efficiency is a bit worn. Such a file will only remove some of the excess metal and will burnish over the remaining metal so that it is flush with the barrel wall. The

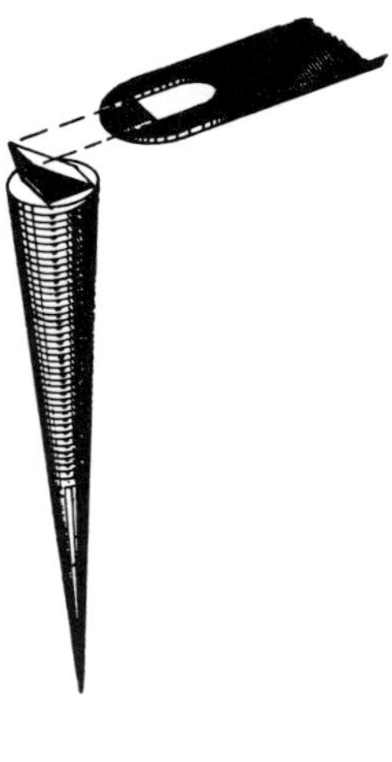

Fig. 11. If hook is for a hole-end spring, it must be especially shaped to fit the hole in the mainspring's end.

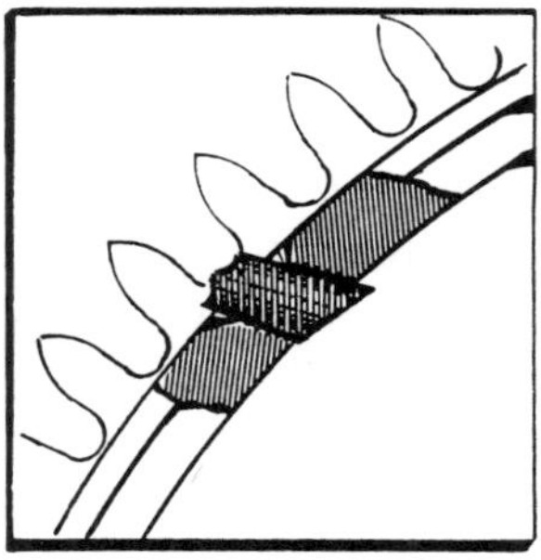

Fig. 12. After taper is inserted and tightened securely, cut off excess (not too close to the barrel wall).

burnished metal will fill up the countersink made by the drill when it started the hole radially. This will secure the hook from both sides of the barrel wall as illustrated in Fig. 13.

The spring is then replaced with a mainspring winder and tested. It should be pointed out that the outside of the barrel wall must present a smooth surface and the burnished-over pin must follow the round shape of the barrel. Otherwise, if the outside of the taper is permitted to appear as a bump, this will stop the watch when it comes into the pitch-path of the center pinion.

REPLACING THE HOOK IN A BARREL ARBOR

When the hook in a barrel arbor is worn, it most often can be dressed up with a triangular or a sharp-edged needle file. Sometimes the hook is worn so much that it is impossible to shape up the hook from the metal around the arbor. The circumference of barrel arbors are usually shaped in a partial spiral so that the contour of the wound spring may be coiled around the arbor, following its natural spiral without kinking or marring the spring. Therefore, to file the arbor excessively to form a new hook would cause spring breakage as well as providing a poor hook at best.

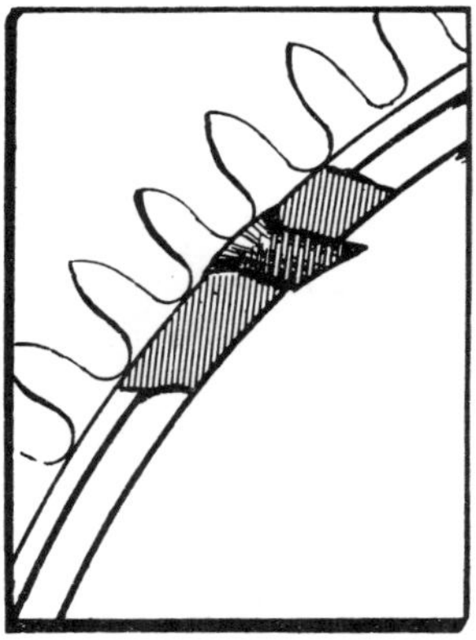

Fig. 13. By burnishing excess with worn file countersink is filled, offering greater security for the hook.

Replacing a new hook is done by filing off the old one and drilling a hole diagonally as shown in Fig. 14. Most barrel arbors are quite hard and it may be necessary to draw the temper from it in order to be able to drill into the metal. Since a barrel arbor has more polished bearing surfaces than any other part of the watch, these polished surfaces must not be burnt by heating the arbor over a direct flame. It can be sufficiently annealed without losing its color by placing it in a pan of fine brass filings and by covering it with

Fig. 14. After annealing barrel arbor, drill diagonal hole into its hub.

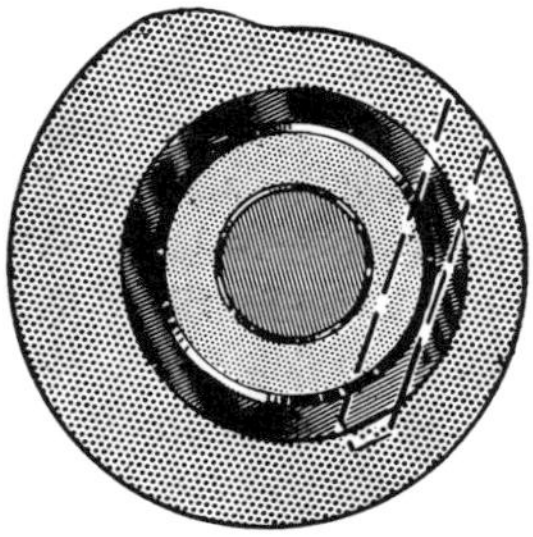

the filings so that its surface will not become oxidized during the heating process. To observe the correct annealing color, a polished ratchet wheel screw is pressed into the surface of the filings so that its polishd head is exposed. The pan is then heated until the polished screwhead turns to a deep blue, which will indicate that the buried arbor has reached the same heat. The pan is then removed from the flame and the arbor taken out. The original polish of the arbor will not be affected. The diagonal hole is drilled with a drill whose thickness is just a bit smaller than the rim-edge of the arbor.

The hole in the arbor should be threaded before threading in the screw which will become the hook. This is best done by threading a long screw and filing off its front end to a tri-cornered taper, the making of which has been previously described in the chapter of screws.

The tapered screw-top is then made to tap the hole in the barrel arbor, using oil to facilitate the process. When the hole is threaded,

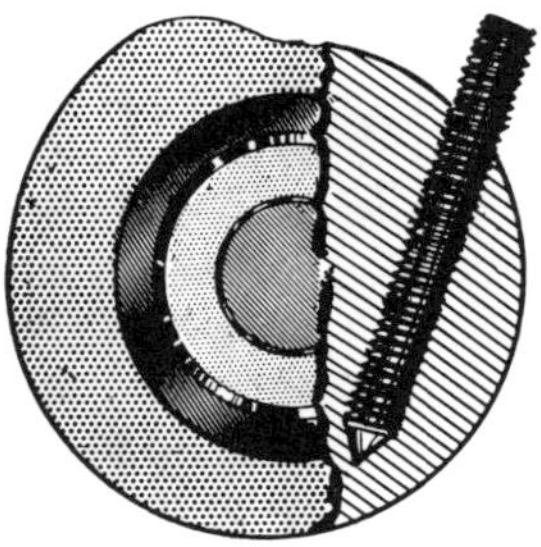

Fig. 15. After hole is threaded, turn plug screw tightly into hole.

the tapered end of the screw is filed away and the rear part of this screw is threaded into the hole until it becomes tight, as shown in Fig. 15. After the screw is absolutely secure, the hook is shaped as shown in Fig. 16. Here it will be seen that the length of the hook reaches up to the dotted lines representing the true (circular) diameter of the barrel arbor. Notice also that if the hook has been drilled and threaded diagonally, the angle will be sufficient to grasp the spring securely.

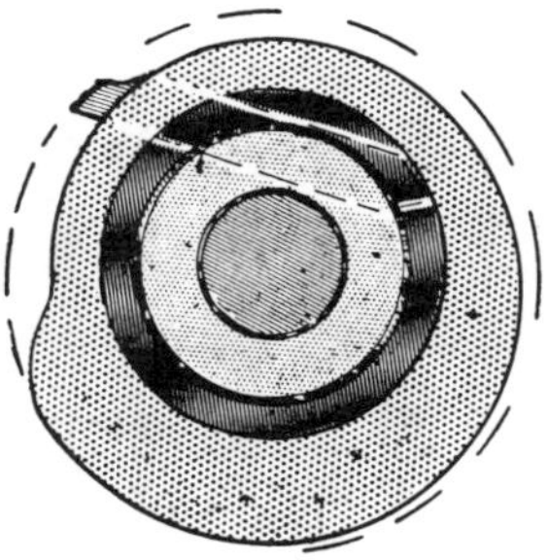

Fig. 16. Length of new barrel arbor hook should not exceed circumference of barrel arbor (dotted circle).

The width of the hook is filed so that it will fit loosely into the hole in the inner coil of the mainspring as shown in Fig. 17. It is very important that this hook should not be left too long as it will kink the coil coming into contact with this hook and every other coil over it. As shown in Fig. 18, this will give a very inferior performance and soon cause the mainspring to break.

Fig. 17. Thickness of the hook must be shaped to fit easily into the hole of the inner terminal of the mainspring; if it is made too thin, it will either break off or cause the mainspring to break at the hole.

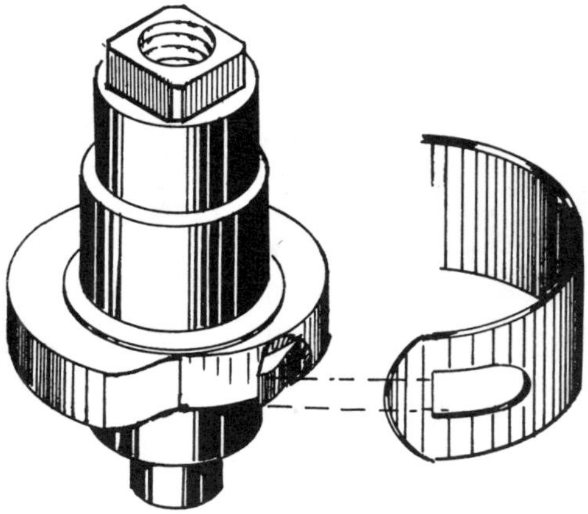

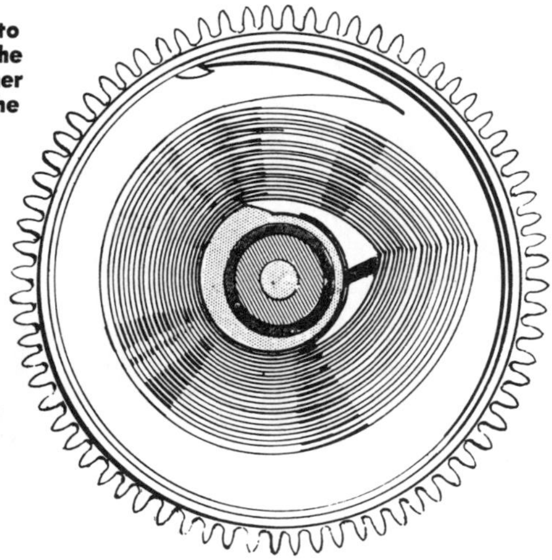

Fig. 18. If the barrel arbor hook is too long, it will kink every coil wound over it and cause breakage.

BOOK III

CHAPTER III

Part I

REPLACING PIVOTS

PIVOTING, one of the finest of watchmaking skills, is the art of replacing a broken pivot by drilling a hole lengthwise into an arbor (Fig. 1)—resorted to when a factory replacement is not obtainable or too expensive, or time does not permit a long wait for a replacement wheel, arbor, or staff. Pivoting is regarded as a standard practice by the best watchmakers, but to be satisfactory

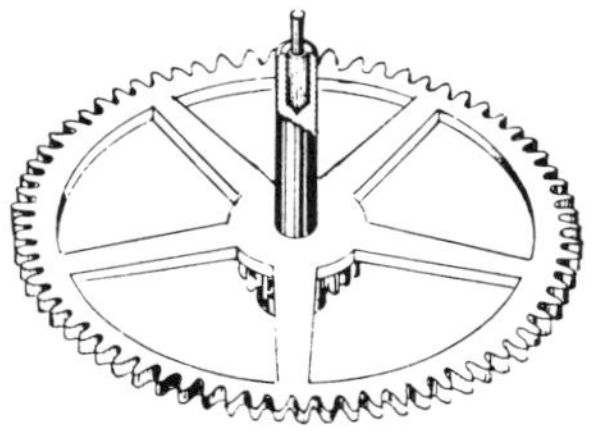

Fig. 1. Arbor repaired with new pivot-plug placed into a hole drilled lengthwise into the arbor.

must be done properly. Operations in pivoting are: (1) spotting the center, (2) drilling the pivot-plug hole, (3) preparing and fitting the plug, (4) finishing the pivot. Each of these operations requires skill developed through practice.

Watch manufacturers, to reduce the weight of train wheel arbors and their frictional surfaces, make the arbors very thin; consequently, any holes drilled lengthwise into them must be centered exactly. A hole drilled off-center weakens that side of the thin wall and increases the possibility of it breaking through and ruining the arbor.

Many watchmakers claim that "catching the center" of the arbor is three-fourths of the pivoting job. There are two methods of spotting the center: (1) using the sharp point of the graver, which is guided by the eye to the center while the arbor is turned in the lathe, and (2) employing centering devices especially designed for this operation. Guiding the graver by the eye requires great skill and practice. In both methods, a centerpoint must be cut into the hard metal post in the place formerly occupied by the pivot.

ANNEALING

Before going further into center-spotting, let us consider methods of annealing (the process by which a metal is made softer by heating and cooling slowly.) Some arbors or posts are so hard that they are not easily cut by a drill or graver. In such cases, soften the arbor only as much as is needed to permit the drill or graver to cut into it. Do not anneal, however, unless it is absolutely necessary because an arbor should be as hard as possible so that it will grip a new pivot firmly as well as maintain the arbor's wearing qualities.

Because there is a danger of over-annealing the arbor and softening adjacent parts, never apply heat directly to the part to be annealed; instead, apply heat by an annealing tool. Heat traveling from the heated tool warms the arbor slowly until the correct

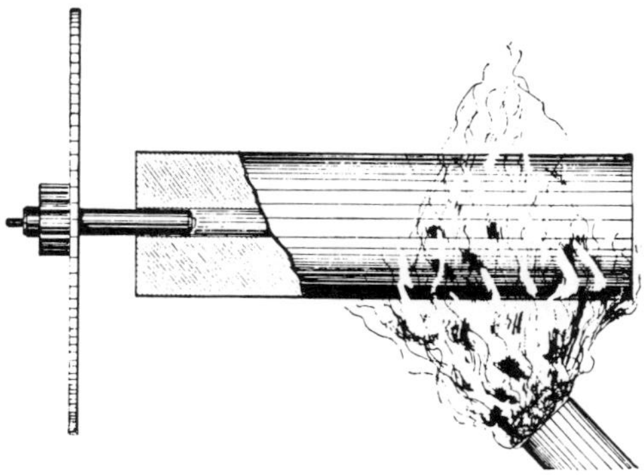

Fig. 2. A copper rod, drilled to fit over the arbor, is heated and used to anneal the arbor.

degree of heat is attained. Annealing tools which may be used are a copper rod, a commercial tool, or special pliers.

A copper rod, into which a hole has been drilled lengthwise so that it fits snugly over the arbor-post, is often used. Heat the rod until the spot on the arbor which is closest to the copper rod turns a dark-blue; then plunge the arbor into water to arrest the color and to prevent the heat from going down further into the pinion (Fig. 2).

A commercial annealing tool (Fig. 3) grips the arbor with a spring clamp while the flame is directed on the wing of the tool.

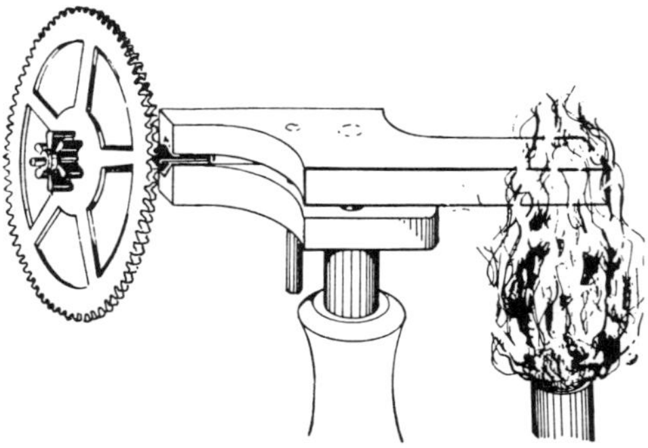

Fig. 3. A commercial annealing tool used to soften the arbor in preparation for drilling.

For annealing, some watchmakers use a pair of special pliers, the jaws of which are heated, after which the arbor is grasped in the jaws, transferring heat to the part to be annealed. Each jaw is lined with a thick piece of copper for better heat induction and has three different-sized grooves so that arbor of various diameters

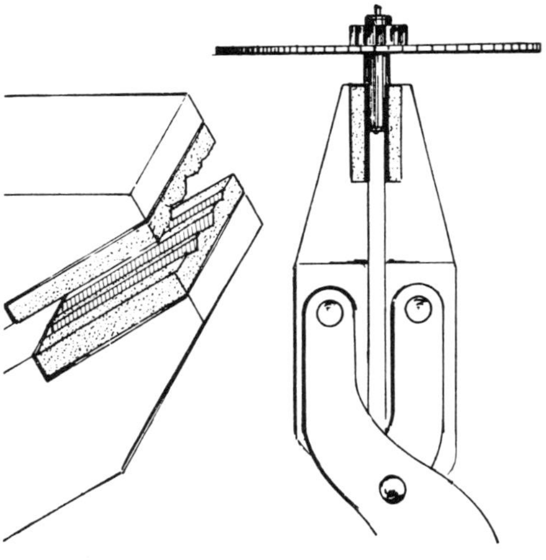

Fig. 4. Brass-lined pliers are heated and then used to grasp the arbor in the annealing process.

may have greater surface contact with the jaws. Such a plier is shown in Fig. 4.

When annealing a train wheel arbor, heat to a dark blue. Do not allow this color to reach the brass wheel or pinion leaves.

When annealing a balance staff, be careful so that no part of the balance wheel is heated to a change of color because the balance wheel would lose its temperature-compensation qualities.

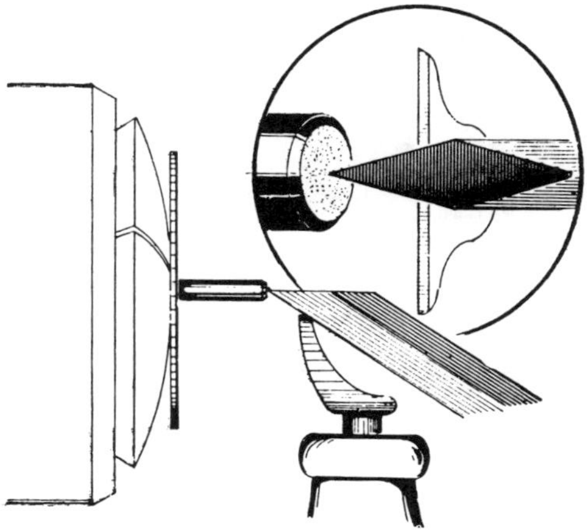

Fig. 5. T-rest and graver are adjusted to strike center of arbor end.

Returning to the job of spotting the center, the pivot hole in the center of the arbor calls for special skill. There are tools which make this job easier. These are called *pivoting lathes* or *centering tools*.

Watchmakers skilled in this art merely fit the pinion or arbor into an accurate chuck, adjusting the T-rest at right angles to the headstock. After facing off the front of the arbor to remove traces of the broken pivot and presenting a clear flat surface, a

Fig. 6. Left, correctly centered cut; right, off-centered cut.

sharp graver is brought cautiously up to the slowly turning lathe and the graver point is carefully made to touch the center of the arbor as shown in Fig. 5. The center can be found by watching the arbor-end through a 10-power (pivot) loupe; the concentric circles forming a kind of bull's eye. The graver is made to touch this center very lightly. The slight countersink thus made is the guide for the drill. A cross section of the arbor should reveal a center-sink as shown in Fig. 6.

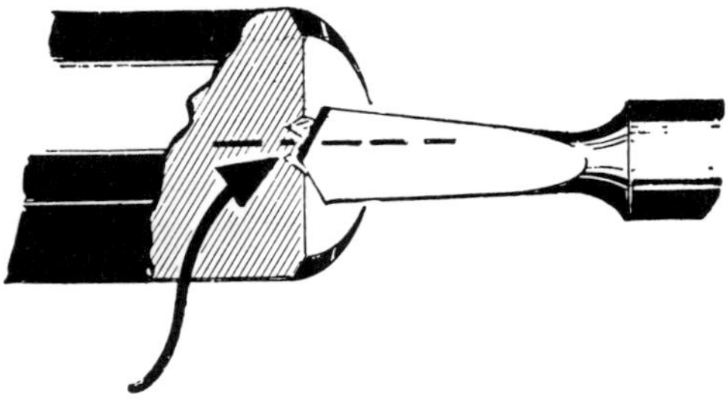

Fig. 7. Drill will cut off-center when true centerpoint is not obtained.

If the graver did not exactly catch the center, a cross-section will reveal this as shown in Fig. 6A, where instead of a countersink, a point surrounded by a circular groove will appear. If a drill is brought into such a cut, it will drill off-center as indicated in Fig. 7.

To correct an error of a countersink cut off-center, the graver is brought to the center of this ring on the front of the arbor, this time more cautiously, making an effort to place the point in the center of the ring previously cut so that a true countersink results.

PIVOTING LATHES AND CENTERING TOOLS

Because of the difficulty in striking a true center, pivoting lathes, sometimes called centering tools are used. One such tool is shown in Fig. 8. This consists of a brass frame "A," which is about 1¾ inches long. This has a step which can be fitted into a benchvise. The lathe frame is drilled through both wings to accommodate the two centers "B" and "C". "B" is the dead center into which the pivot of the arbor opposite to the one to be replaced is rested. The end to be drilled is placed into the hollow center of spindle "C". A pulley or ferrule "D" fitting over the spindle "B" has two pin projections which are made to straddle a spoke of the wheel to be pivoted.

The ferrule is rotated by a small bow and string "E". A pivot drill "F" in a holder "G" which closely fits the hollow section of spindle-center "C," is placed through the back of the spindle. This reaches through until it contacts the flat-front end of the arbor which is resting in the mouth-tapered end of the hollow spindle. The stationary drill cuts the hole in the arbor as the wheel is rotated by the ferrule. Compartment "H," covered by screw "J," holds extra drills. Some watchmakers use these little tools to drill the entire plug-hole while others merely use it to obtain the center-spot and then continue the operation in the bench-lathe, using regular chucks and pivot-drills.

While pivoting lathes such as pictured in Fig. 8 may still be purchased, for the many watchmakers who do not possess these, a simple method of spotting the center is recommended. This method

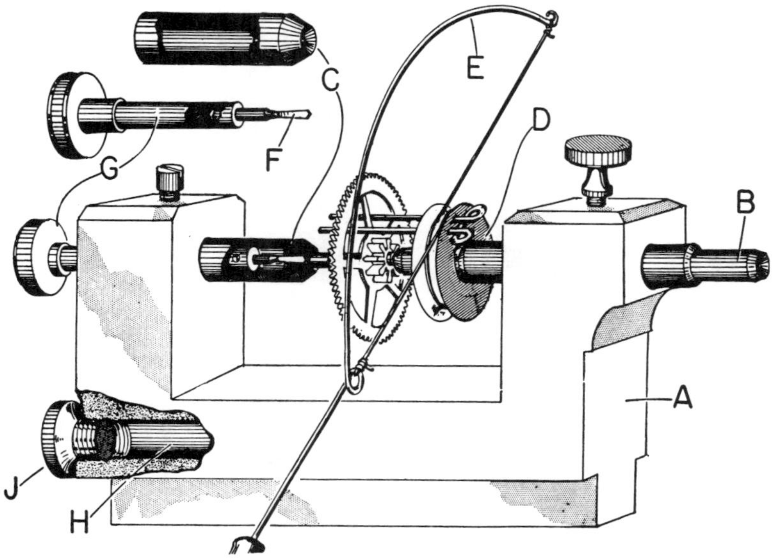

Fig. 8. A pivoting lathe and accessories with wheel ready for drilling, shown in use. (Parts A, B, C, D, E, F, G, H explained in text.)

consists of the use of a length of mainspring from a 12-size watch and about two inches long. This is softened at one end and this is placed on a soft piece of brass or aluminum which rests on the die of a staking set. Place a well-conditioned pointed punch over the softened end of the spring and give it a sharp blow. This will

Fig. 9 Fig. 9A

Fig. 9. A strip of mainspring indented with cone punch is prepared for job as centering device.

indent a cone-shaped depression as shown in Fig. 9 and in cross-section view 9A.

Next take a pivot drill about .15 mm. and drill through the depression as shown in Fig. 10 and 10A. Drilling through the exact center is simply done by placing the drill into the lowest part of the depression and permitting the drill to follow its own course. To spot the center of the arbor to be drilled, the wheel is placed in the lathe and the coned depression is placed over the end of the arbor which has been made flat on its end, slightly chamfered to fit the cone-depression. The spring is held by the left hand against the arbor-end at right angles to the axis of the arbor. In the right hand, the drill which was used to drill through the

Fig. 10 Fig. 10A

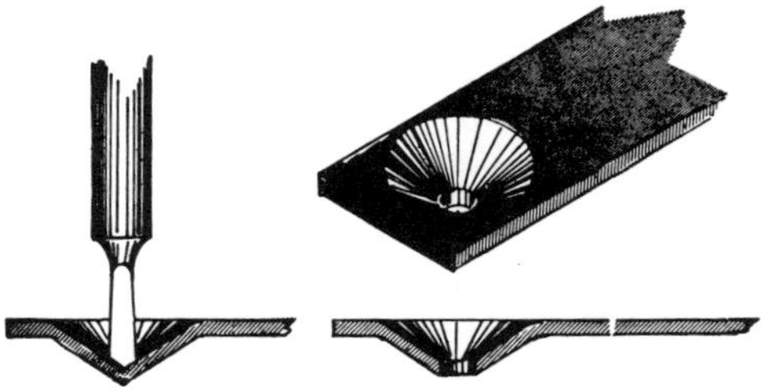

Fig. 10. Drilling through the cone depression to make centering device.

cone-depression, is placed through the rear of the cone which is now over the arbor.

The drill held in a light pinvise is then against the exact center of the arbor and a few turns of the lathe is sufficient to center-spot the arbor-end. Since the end of the arbor fits into the coned-depression of the spring it centers itself. Because the hole in the spring's cone is drilled exactly in center, the drill guided by the hole it previously cut will also spot-drill a hole exactly in center. This is shown in Fig. 11.

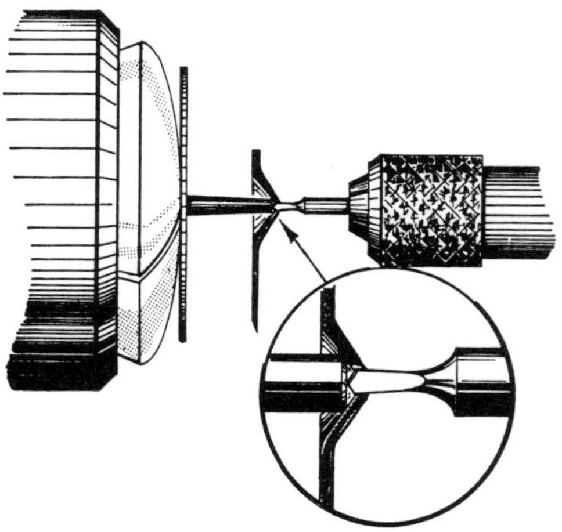

Fig. 11. The same drill is used to make the hole in the spring's depression, and is used through reverse side to strike the centerpoint in the arbor-end.

DRILLING PIVOT HOLE IN PINION

Drilling the hole in the arbor end calls for care as well as a choice of drills. While it is possible to obtain pivot drills in sizes from .10 mm. (or .004 inch) and upwards, it is highly advisable that every watchmaker possess the skill necessary to make his own pivot drills. Commercial drills often need re-sharpening and servicing; such work involves practically all the operations of making a pivot drill.

Knowledge of kinds of steel to be used, shaping the drill-blank, filing it to shape, tempering, and sharpening are essential in making pivot drills. Fig. 12 shows the type of drill in use today. It is simple in construction and easy to duplicate.

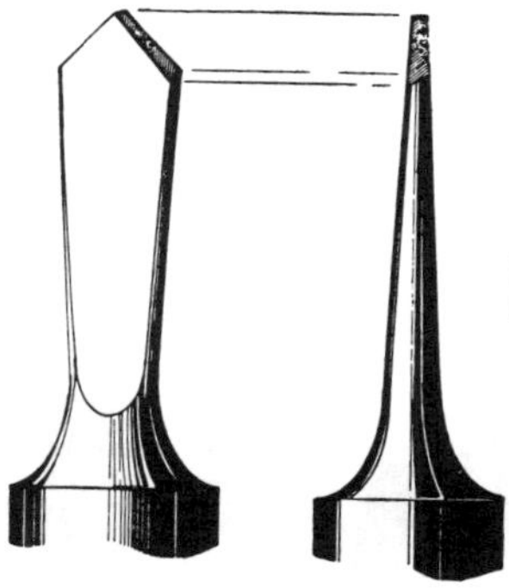

Fig. 12. Two views of modern type pivot drill.

To make such a drill, use high carbon steel or drill rod. Hold the soft steel in the chuck and turn it to shape (as shown in Fig. 13). The blank should be turned so that it has a slight back sloping angle ("A") to supply a clearance for the drill and room for metal shavings and chips.

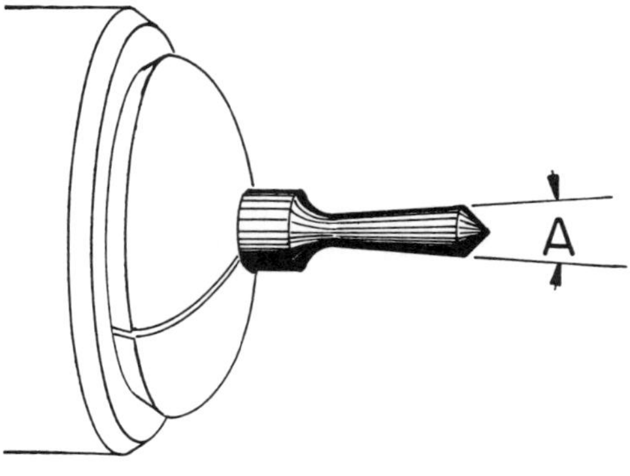

Fig. 13. The drill blank is turned to the shape indicated. Carefully observe the required back taper.

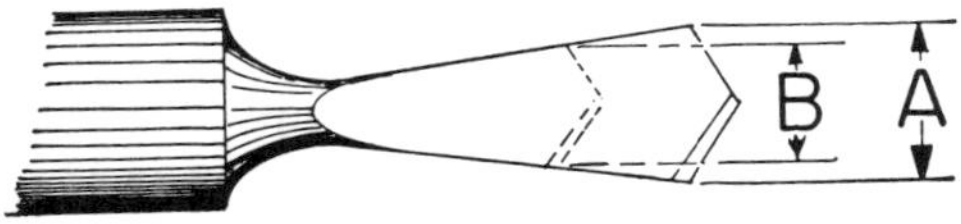

Fig. 14. If the taper is cut back too much, the drill is weakened and subsequent sharpening drastically reduces its width.

If this taper is cut back too far (Fig. 14), the drill will be weakened and the possibility of breaking increased. Furthermore, every time the drill is sharpened, it will not only become shorter but thinner. Be certain not to make the drill too long; rather, make it just a bit longer than the full depth of the planned pivot hole, allowing extra length for future sharpening.

After the back taper has been cut, the drill is filed flat on both sides. The flat surfaces should be tapered towards the drill's tip as shown in Fig. 15. These tapered flats are filed with a fine file while the drill blank is still in the lathe. For this operation, the

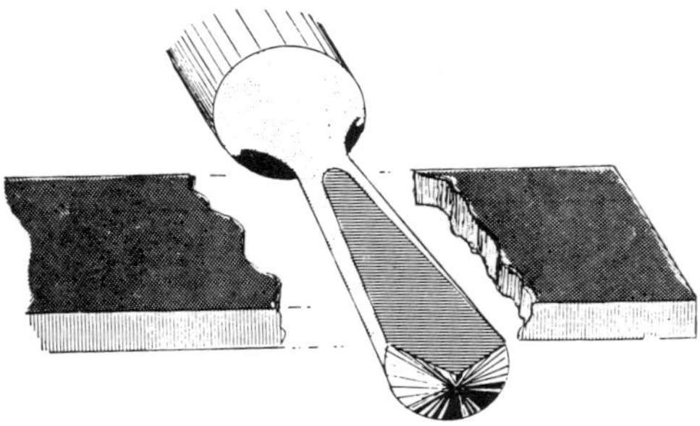

Fig. 15. When flat filing the drill, the flat surface should end at the small section of the shank.

lathe is locked by the locking pin in one of the quarter holes of the index ring at the rear of the lathe pulley. In order to guard against filing the flat too low, frequent observation should be made of the tip-point on the drill blank. This serves as a guide against over-filing. The flats must be equidistant from this point.

In filing the tapered flat, it will be noticed that this blends into the rounded rear section of the shank. Care must be taken that these flats are also of equal length. If the flats are filed to unequal

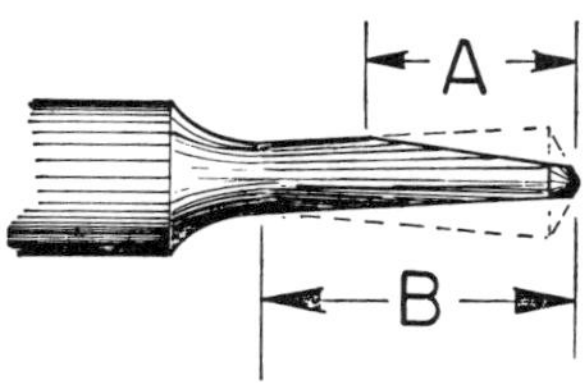

Fig. 16. Taper "A" is too short; taper "B" is correct. A drill with unequal tapers will cut off center.

lengths as shown in Fig. 16, the drill will cut off-center. If they are of equal length, it is additional assurance that the drill's two flat surfaces end equally distant from the front tip.

After one side has been filed, the index pin is withdrawn and the lathe head turned half-way around and the pin re-inserted so that the filing may be completed with both sides opposite and parallel. The file is carefully manipulated so that each stroke brings the taper a little further back from the tip until it blends into the round section opposite the taper-end on the other side. The thickness of the blade should be about 1/3 to ¼ its greatest width as shown in Fig. 17.

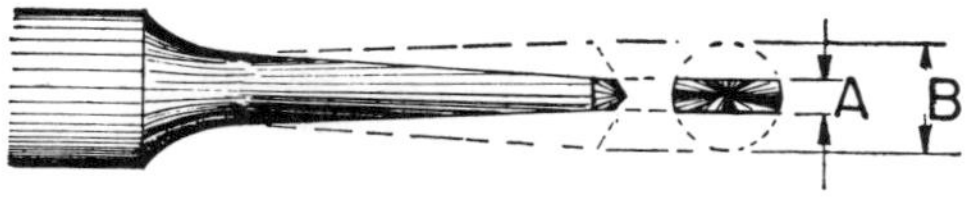

Fig. 17. Thickness of blade tip "A" is generally one-third of its width "B" in smaller drills, and one-half of "B" in thicker drills.

Filing the flats may bend the annealed blank slightly but when the opposite side is filed, it will straighten it somewhat. However the drill should now be checked for truth by turning it slowly in the lathe and observing its tip-point, bending it carefully until it is again on-center. It is then removed from the lathe after which it is hardened and tempered.

TEMPERING THE DRILL

Tempering the drill is done by holding it over an alcohol lamp flame with a heavy duty tweezer at the end farthest back from its point with a jar of water very close by. When the drill becomes red-hot, quickly remove it from the flame and plunge it *vertically* into the water as quickly as possible. Plunging it into the water vertically removes the possibility of warping the drill as it is quenched. This makes the drill dead-hard and too brittle for use. Therefore the drill should be tempered by annealing it enough to remove the brittleness yet retaining sufficient hardness to cut a reasonably hard piece of steel.

Annealing the drill to complete the tempering process is done in a different manner than with larger and heavier pieces. Some watchmakers prefer to draw the temper to a yellow straw color for average use, but for pivoting in steel which is still tough although annealed, the following process is recommended. This process is based on the thought and practice that if it were possible to temper the drill so that its point were very hard, but its shank softer and springy, the drill would cut with the stress or strain being absorbed by the more resilient rear of the drill.

To achieve this result in tempering, the flat surfaces are ground on an oilstone so that the file marks are removed. This sharpens its edges and provides a clean surface for observing the color changes during the tempering process. This can be done without breaking the brittle drill by placing the flat part of the drill over the edge of an oilstone slip with the soft tip of the index finger

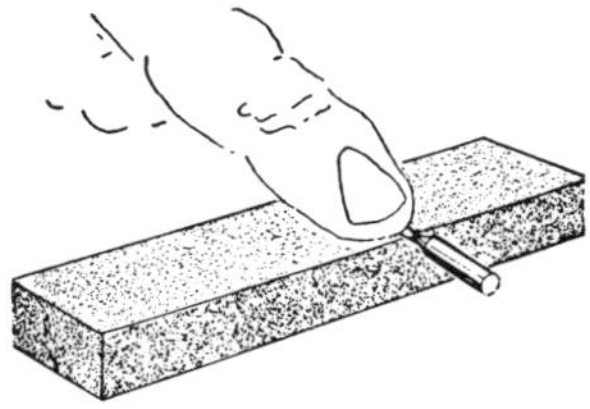

Fig. 18. Flat sides of drill may be brightened for observing tempering colors by moving the flat part of the drill across edge of oilstone.

over it as shown in Fig. 18. Moving this carefully back and forth a few times will remove the file marks and bring up a bright surface which is better for observing the tempering color. Next the very tip of the drill is grasped by a tweezer with heavy blunt ends such as used for straightening pivots. Then the heavy rear part of the

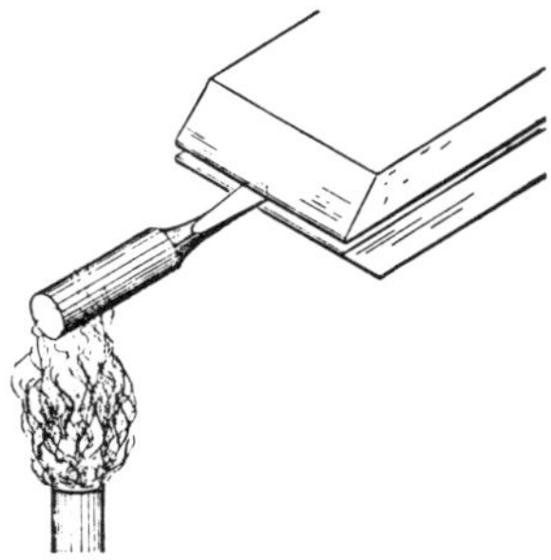

Fig. 19. Heating shank of drill while tip is held in heavy-duty tweezers gives better control of tempering process.

drill is held high above the flame of an alcohol lamp as shown in Fig. 19. By carefully moving this part of the drill back and forth across the flame, the color changes will slowly creep from the heavy part of the drill towards the part held by the tweezers. The darkest color will first show up in the rear of the drill and the lighter colors, advancing before it will move towards the tip of the drill. When a light straw color reaches the tip of the drill held in the tweezers, quickly plunge the drill vertically into the water to arrest any further annealing of the drill. The tweezer, holding the drill-tip absorbs any extra heat, preventing softening the drill

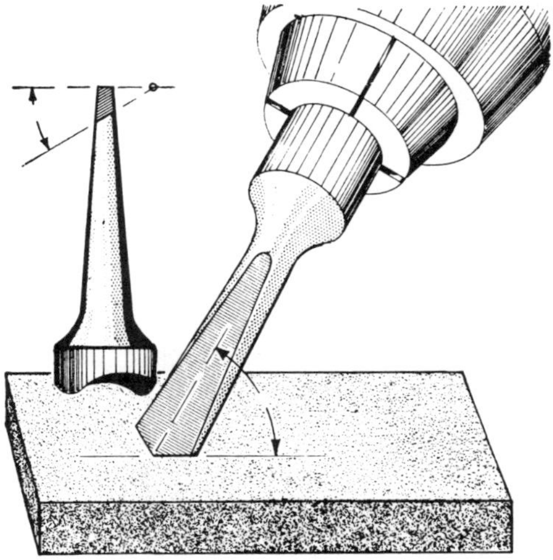

Fig. 20. Sharpen drill by tilting it backwards and stroking across oilstone. Cutting edges are made so that drill cuts with clockwise turning. Upper left figure shows proper angle for cutting edges.

tip while the edges near it are still tough and springy. It should be emphasized that the rear taper of the drill should not be allowed to be heated more than deep blue color (after which the color becomes paler and the metal too soft).

The drill is now ready for the finishing touches which will provide the cutting rake to the lips of the drill. For this job, the drill is held in a light pinvise. The lip should be made to contact the oilstone so that no light can be seen between it and the oilstone from its point to the side edge. Then to provide the cutting edge, it is tilted slightly without spoiling this contact. This is shown in Fig. 20. In this position, the drill is rubbed lightly across the stone in

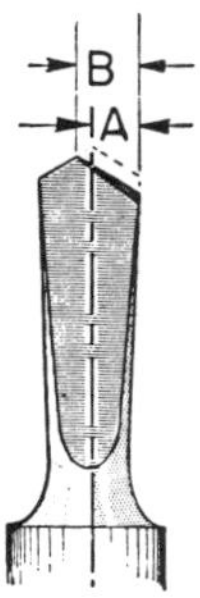

Fig. 21. If the drill is sharpened unevenly, it will cut too large a hole. "A" shows center of the drill, and "B" the radius of hole this drill will cut.

short strokes to assure a sharp flat cutting rake. The drill is then turned around and the other half sharpened. Make sure that the cutting edges meet each other exactly on center, guided to this by the original center point. If one side is sharpened more than the other, the cutting point will be off-center (Fig. 21) and any hole it will cut will be much larger in diameter than the hole desired, with the added danger of breaking through the walls of the arbor. When sharpening drills this must always be remembered.

The hole in the arbor should always be a little greater than the diameter of the pivot to be made, so that any inaccuracies in drilling the hole can be corrected when finishing the pivot.

In drilling the hole in the arbor, hold the pinion and arbor in an accurate chuck; and drill the pivot hole while holding the drill in a pinvise. Start the drill in the spot-center.

PART II

FITTING THE PIVOT

WHEN an upper escape wheel pivot is broken, or a pivot atop some other long-pinioned wheel must be replaced, prepare a cement chuck and cement the wheel to it by shellac because the lower part of the arbor is too short for a chuck to grasp it.

In this case the cement chuck is a rod of brass. A countersink is cut shallow enough into it so that the arbor pivot on the

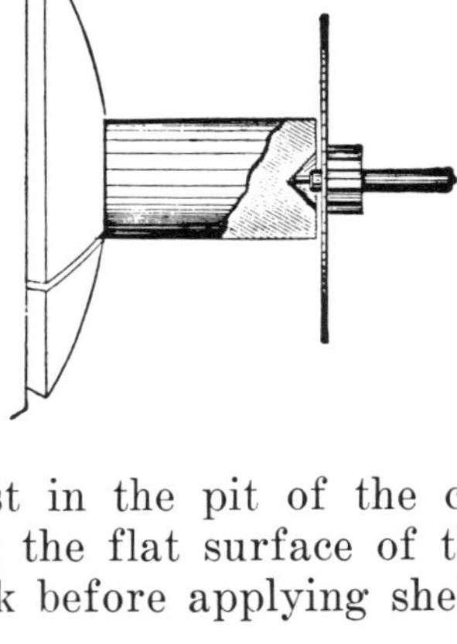

Fig. 22. Cement brass is prepared for pivoting wheels difficult to hold in regular chuck.

opposite end may rest in the pit of the countersink without the escape wheel touching the flat surface of the brass rod (Fig. 22). Heat the cement chuck before applying shellac.

While the shellac is still soft, place the wheel in the semi-liquid mass so that the back pivot rests in the countersink. The back pivot can be kept in the base of the countersink of the cement

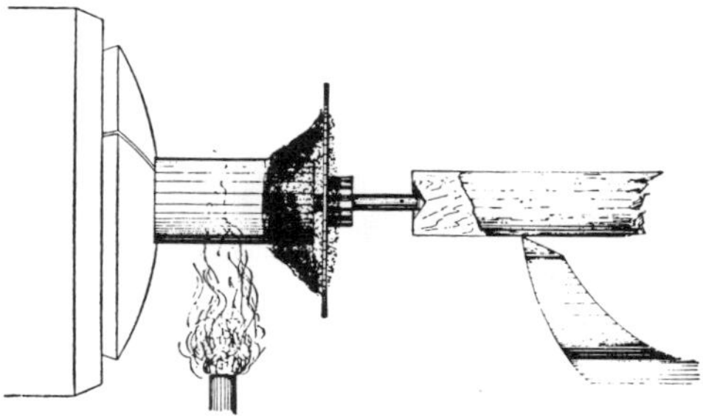

Fig. 23. Hold wheel in cement chuck with shellac, and true with peg-wood end.

chuck by pressing on the front of the arbor with a piece of peg-wood resting on a T-rest, its end countersunk. The wheel can be perfectly trued while the cement chuck is kept turning in the lathe, using the pegwood to support the arbor until the shellac has hardened (Fig. 23).

After finishing the replaced pivot, dissolve the shellac with alcohol.

The arbor hole for pivots of ordinary length should be at least one and one-half times as deep as the length of the pivot it will replace. Although a deeper hole may furnish a greater gripping surface, the chances of breaking a drill while making such a hole are greatly increased.

When drilling the hole (Fig. 24), grasp the pinvise with moderate pressure so that the drilling can be "felt" by the fingers holding the pinvise. A little light oil will help the drilling. The drill should

Fig. 24. Hold pivot drill (in light pinvise) in the hand for the delicate drilling operation.

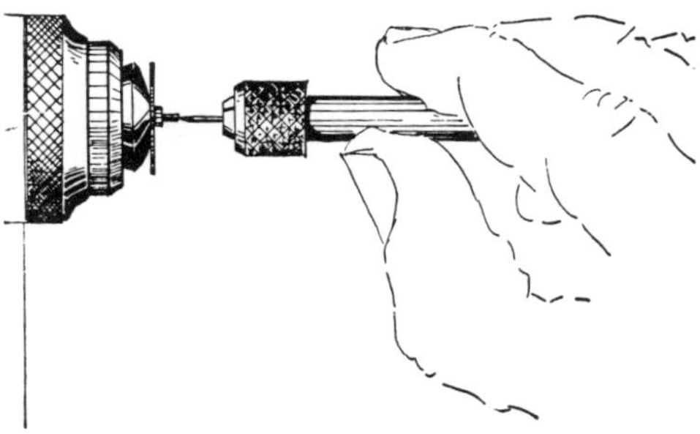

be removed frequently so that chips may be cleared from the hole to prevent clogging and breaking the drill in the hole.

Sometimes during the drilling operation the drill will stop cutting. Even increased pressure will not cause the drill to cut. This is the result of turning the work too fast—the thin drill becomes hot; the heat softens it and dulls its cutting edge and burnishes the inside of the hole, making it still more difficult to cut with ordinary drills.

This burnished surface can be removed by either re-annealing the arbor or using a flat chisel-shaped drill (Fig. 25). The cutting edges of such a drill attack the weaker edge of the burnished hole, removing the glossy surface as it reaches the bottom. Resume cutting with a hardened, tempered, and re-sharpened pointed drill until the desired depth is reached.

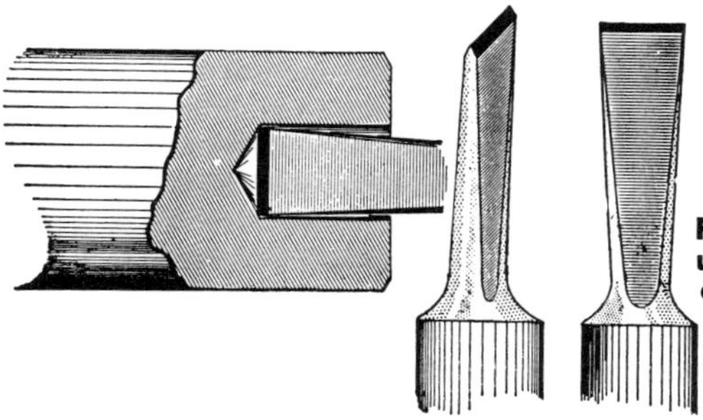

Fig. 25. When bottom of hole becomes glaz use a chisel-shaped drill to cut away the gla attacking the outside edges and then working wards the center.

Occasionally, a drill will break in the hole just as the desired depth is reached. This generally is the fault of either drilling too far, exerting too much pressure on the drill, holding the drill off-center, or because the drill has some inherent weakness.

When the drill breaks, it "grabs" the metal and becomes imbedded in it. If the pieces of drill cannot be removed, it is still possible to complete the job if the arbor is annealed again, softening the broken drill as well as the arbor. If the imbedded piece of broken drill is relatively small, drill the hole carefully again; almost always the broken drill will come out as shavings.

PREPARING THE (PIVOT) PLUG

Metal used for the pivot plug should be of a good grade steel, hardened and tempered to a blue color.

Turn the pivot plug in the lathe so that it is barely tapered, and make the end thin enough so that the top of the pivot enters the hole.

The spot on the taper, marked where it entered the hole, indicates the hole size.

Grind the plug *cylindrically* with an oilstone slip until the plug goes into the hole almost one-third of the depth of the hole.

Turn a light groove on the plug with a graver (Fig. 26) a little

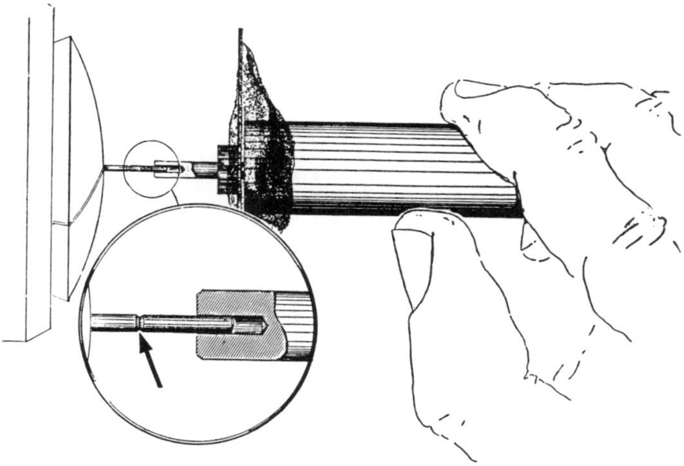

Fig. 26. Oilstone powder helps to work pivot plug into hole. Groove (arrow), a safeguard, furnishes automatic break-off point when plug "freezes." If wheel is in shellac chuck, do not remove it until the entire pivoting job is completed.

above the intended length of the pivot being made. While the plug is still in the lathe, coat the pivot plug with a light mixture of oilstone powders and oil. Hold the wheel in the fingers (or if still shellacked) in the cement brass.

Press the wheel over the slowly turning plug, which will grind itself into the hole, aided by the coating of oilstone mixture.

When the plug reaches a depth in the hole where it is gripped tightly, it will suddenly freeze. At this instant, the strain on the plug will cause it to break off at the groove (which marks the length of the pivot). The pivot and arbor will be secure, allowing

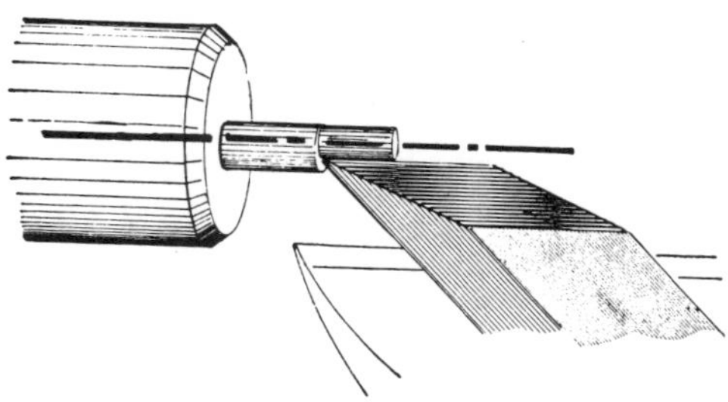

Fig. 27. Pivot plug should purposely be made a little thicker than required for finished job. This permits truing in event hole was drilled off-center.

the pivot to be trued by carefully turning it in the lathe and trimming off out-of-round portions (Fig. 27).

Some watchmakers insert the new pivot plug by tapping it lightly with a hammer. While this procedure may do the job, it often will cause the pivot to bend or place a strain on the walls of the arbor plug hole.

FINISHING THE PIVOT

When finishing straight pivots with square shoulders for a wheel arbor, examine the base of the pivot to make certain that it has no radius which might become wedged in the jewel hole (Fig. 28, "A").

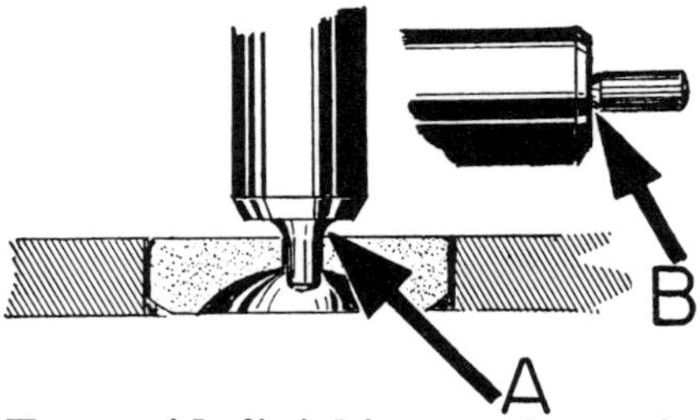

Fig. 28. Pivot-corners should be square; oth- wise radius corner will hang on jewel-hole edg "A". Slightly undercutting "B" insures square shou der after grinding and polishing.

To avoid finishing such a pivot, it is advisable to make a slight undercut (Fig. 28, "B") so that when the pivot is ground and polished to size the corner will finish up square.

Some types of pivoting jobs call for special consideration and procedures. Among these are pivots for balance staffs, pivots for (fourth wheel) carrying the second hand, pivots for wheels of ultra-thin watches, and pivots for center wheels.

PIVOTING BALANCE STAFFS

Replacing pivots to a balance staff call for the same operations used with wheel arbors except that the front end of the balance staff pivot-stump should not be faced off any lower than is intended for the thickness of the drill hole. Otherwise the replaced

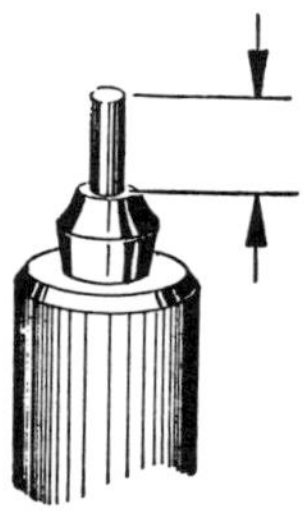

Fig. 29. If too much of the cone is faced off before drilling, pivot will be too long and weaker than one which is correct in length and shape.

pivot will not blend into the general curvature of the cone and look like that in Fig. 29. Such a pivot is necessarily longer and weaker than the one it replaced. Fig. 30 shows the correct amount of cone

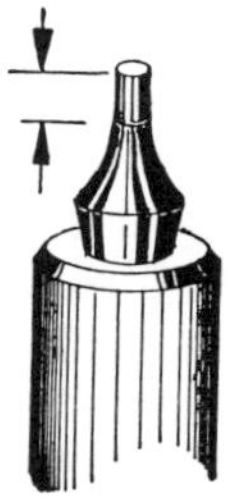

Fig. 30. Pivot correctly faced off.

to be faced off with the replacement.

The pivots for balance staffs are inserted and trued in the same manner as square shoulder wheel pivots but the finishing may be done with an oilstone slip, jasper stone and burnisher, following the curvature of the cone. Too much pressure on the pivots during polishing should be avoided as this might loosen the pivot, although if the directions in grinding in the plug are followed, no such difficulty should be encountered.

PIVOTING FOURTH WHEEL SECOND-HAND PIVOTS

If the second-hand pivot of a fourth wheel has to be replaced, it is always in danger of being pulled out again whenever the second hand is removed from it. Therefore, it is better to drill the hole as wide as is safely possible so that when the pivot is trued, polished and finished, its base will have a shoulder which will brace itself against the hole jewel when the second hand is forced off it. Such a pivoting job is shown in cut-away view in Fig. 31.

PIVOTING WHEELS FOR ULTRA-THIN WATCHES

When a pivot is to be replaced for wheels with very short arbors, it is difficult to adhere to the rule of making the depth of plug-hole 1½ times the height of the pivot. In such a case it is better

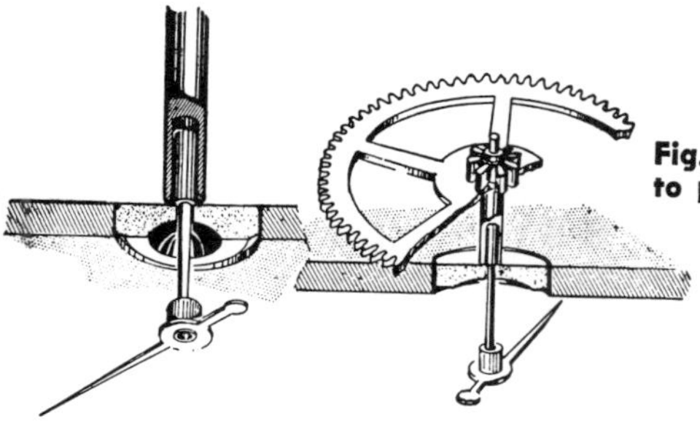

Fig. 31. Have the fourth pivot-plug thick enough to provide a shoulder at the base to prevent its extraction when the second hand is pulled off.

to drill completely through the pinion or arbor. A new pivot post should then be placed friction tight in this hole using the same

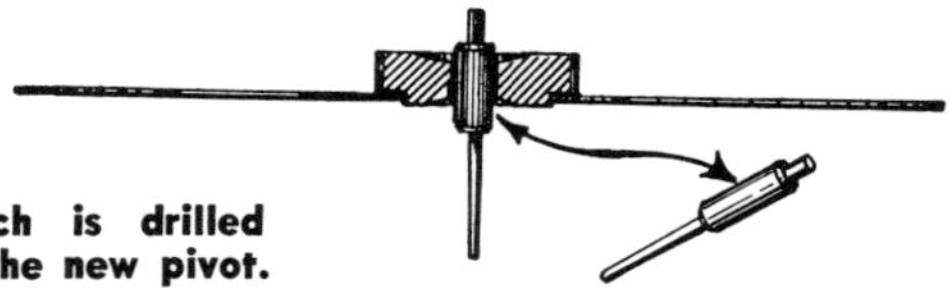

Fig. 32. A pivot for a thin watch is drilled through to provide efficient grip for the new pivot.

methods described in "freezing" the pivot plug in a blind hole, and then finishing the pivots. Or, a new friction post, pivoted to fit the jewels, is placed in the now hollow pinion much like a friction pallet arbor is replaced in a pallet. Such a replacement is shown in Fig. 32.

PIVOTING CENTER WHEELS

By using the method previously described for pivoting short arbors, center wheels with broken posts may be reclaimed. In the case of a center wheel, the job is not quite as difficult but calls for more detailed work. When the center post is broken it usually breaks just at the main plate bearing as shown in Fig. 33.

In such a case, the height and thickness of the shoulders and pivots should be gauged and noted. The center pinion, held in the lathe chuck presents the rear pivot which should be faced off and

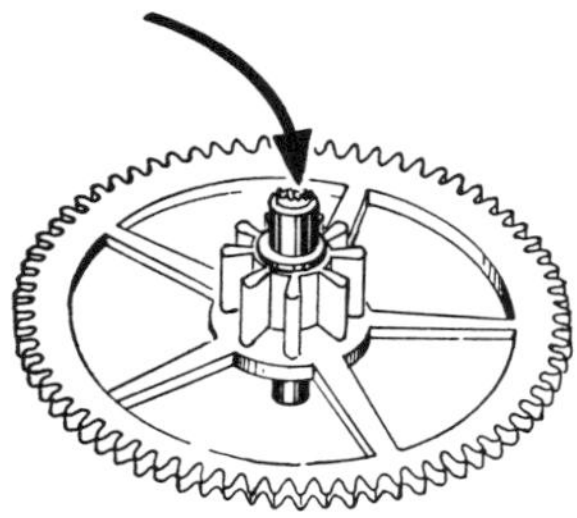

Fig. 33. When a center pinion breaks, it usually occurs at the base of the center-post.

spot centered and then drilled clear through. The diameter of the hole should be as large as possible without weakening the pinion. This often will cause both pivot shoulders to be drilled away. The purpose of drilling the hole of such dimensions is to permit an inserted arbor to be of sufficient diameter to provide for shoulders, pivots and the cannon pinion post.

When the inserted arbor is fitted friction tight into the hollow pinion, its shoulders, resting between the upper and lower center-jewel holes will resist any upward pull when the cannon pinion is to be removed. The same arrangement prevents it from being pushed out of the center-pinion when the cannon pinion or minute hand is pressed down upon it.

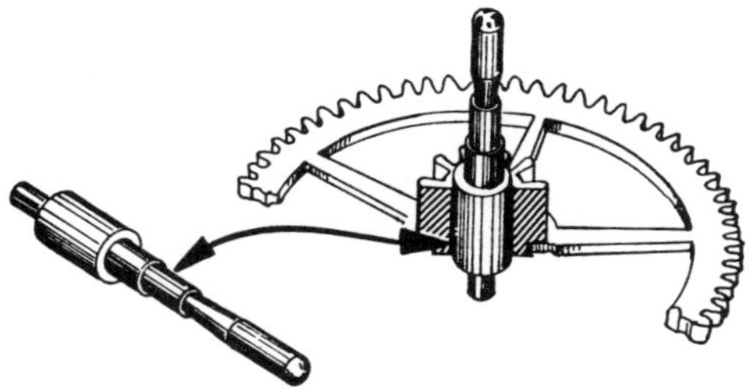

Fig. 34. A broken center pinion may be repaired by drilling through the pinion and making a new post to fit into the drilled out section.

The arbor is turned separately. It is similar in appearance to Fig. 34 and is driven in friction tight into the hole in the pinion. Care should be taken that its position in the pinion hole corresponds exactly to the height required of the center wheel to meet the third pinion correctly, and also to clear the barrel as well as the balance assembly and the underside of the bridges.

The arbor should be tried out for fit in the jewel holes and for the height between plates and cannon pinion groove. All these should be checked before fitting it into the pinion.

Some precautions in drilling the hole in the center pinion should be noted. The hole drilled in the pinion should not be so large that the riveting for the center wheel is cut through and the wheel falls off. For this reason the hole is drilled through from the rear so that it enables one to determine the largest size hole without cutting through the wheel riveting shoulder. Also, the shoulder should be slightly undercut so that when they are finished, the pivot shoulders will be square.

REMOVING TEMPERING COLORS

After the riveting job is completed it will be desired to remove the colors in the polished arbors induced during the tempering process. There are a few chemicals that will remove the colors effectively. One of these is called ''Coeruline''. Another is to boil in a solution of equal parts of hydrochloric acid and water. It is also possible to remove the blue from the steel by dipping the part in ordinary blue ink which contains the necessary chemicals to remove the bluing. The parts should then be washed immediately with soap and water and then cleaned and dried in alcohol and sawdust to neutralize these acids.

BOOK III

CHAPTER IV

DIAL REPAIR AND ADJUSTMENT

No matter how well a watch may be repaired or how polished and clean parts of the movements may be, the dial is the first and often the only part of the watch seen by the wearer—and the workmanship of the watchmaker is often judged by the appearance of the dial. A lack of knowledge of the care and repair of dials causes many repair jobs to be needlessly delayed, spoiled, or made more expensive.

Repainting or refinishing of most dials is not expected of the average watchmaker at the bench, but the skilled watchmaker should be able to select and match a replacement, soft solder a dial foot to any required position, repair chipped porcelain dials, and know the general trade practices in cleaning various types of dial surfaces without damaging them or making them less attractive.

The sections of a watch dial are: the base metal, background finish, bezel opening, numerals, and feet (or other means of attaching). The base metal used for many dials is of brass and copper, although silver is often used for dials of high quality.

The background or color scheme of the dial is generally coordinated with the design or motif of the case. The dial may be of silver, gold, black lacquer, or various shades of enamel. Some dials have a copper base with a porcelain covering and black numerals. These dials are used mostly where a quick and accurate reading is required (such as in railroad watches).

The figures on almost all dials are either painted, stamped or printed, depressed or engraved, embossed, raised or etched. In some cases the numerals are made of high karat gold and applied individually to the dial. Some of these are shown in Fig. 1.

Almost all dials are held in place by feet which fit into holes in the movement and made secure by screws; dials also may be held fast by screws fitting into the movement through the top of the dial. Dial feet are made of copper or brass wire and are hard-

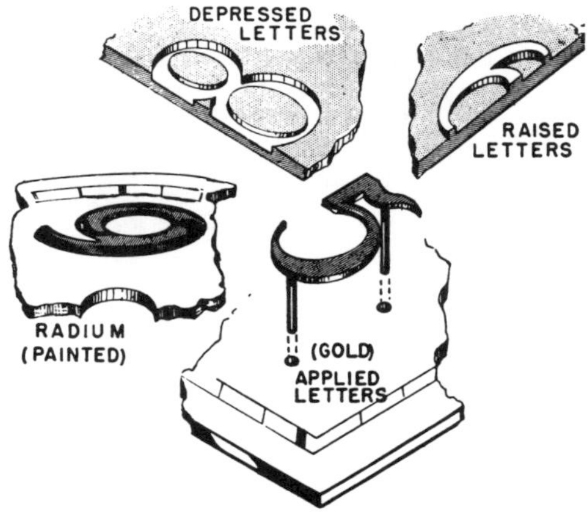

Fig. 1. Types of dial figures.

soldered to the back of the dial before the dial is printed. Some dials made in the factory for a specific model have their feet and base stamped, and extruded from one piece of metal.

MATCHING A DIAL

Occasionally a dial will have to be replaced because of damage beyond repair. A new dial should match the old one in shape, bezel opening, and hour-minute and second-hand hole positions.

In selecting a dial of the proper shape, attention must be directed to the dial opening in the watch case. If the dial has an outside minute track, the entire track must be visible when the dial is placed under the bezel; in addition, the shape of this track must

Fig. 2. Dials with minute tracks must have them conform to bezel shapes and openings.

conform to the shape of the bezel opening and no readable part of the dial should be obscured by the bezel (as shown in Fig. 2).

To compare the replacement with the sample, use a piece of stiff paper about the thickness of a two-cent postcard. The dial feet of the sample are pressed through this cardboard and the dial is brought down flush with the paper. As shown in Fig. 3, with a pencil trace the outline of the dial ("A") and mark the center hole ("B"), second-hand hole ("C"), and the crown position ("D"). The position of the feet is at ("E").

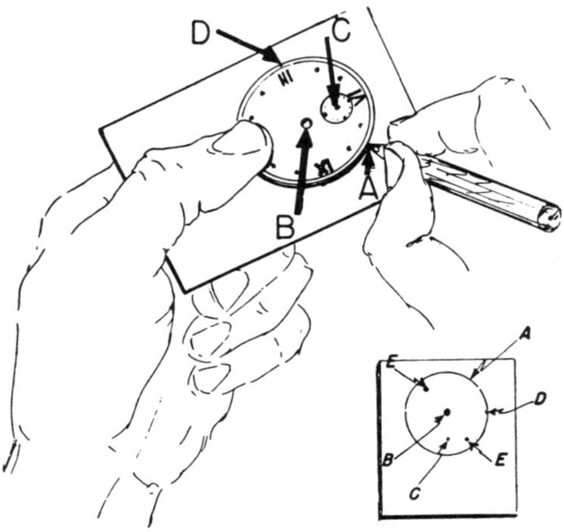

Fig. 3. Method of making a tracing of dial for matching purposes.

Place the new dial on the cardboard and fit the feet of the new dial into the holes made by the feet of the sample. Carefully observe that: (1) the feet fit exactly into the holes, (2) the shape of the dial conforms to the cardboard (tracing), (3) the dial holes and crown position align.

To check the new dial openings for the figures and minute track, place the dial under the bezel of the case and observe the figures and minute track through the crystal. The figures and minute track must not be obstructed and must match the shape of the bezel opening.

FITTING A NEW DIAL FOOT

Special repair jobs which must be delivered on short notice permit no time for refinishing an old dial. In such instances, select a dial similar to the original and having bezel opening and hand-hole positions the same as in the original dial. Remove the feet from the dial selected and solder new feet to it, in the position of the feet on the original dial. This must be done as follows:

With a sharp cutting plier, cut both feet off close to their base and file the stumps of these feet so that the back of the dial is clear and flat. In this operation, caution must be exercised or the face of the dial may become smeared or scratched. To avoid defacing the selected dial, place a piece of tissue paper over its face so that the fingers will not touch it. Another precaution to avoid defacement: do not let the filings fall between the tissue and the dial face. While filing the feet stumps, the dial is held between the forefinger and thumb, the forefinger supporting the dial directly under the stump. Holding the dial in this manner will prevent a small bump on the dial face caused by the file pressing on the dial feet.

After the stumps of the feet have been filed flush with the lower surface of the dial, place the dial over the cardboard so that the hand-holes and crown position are aligned, to mark positions of the dial feet. The dial is pressed firmly against the cardboard with the fingers so that it does not shift (as mentioned previously, the face of the dial is separated from the fingers by tissue paper). With the cardboard on the top, make small indentions on the back of the dial through the holes in the cardboard (which were previously

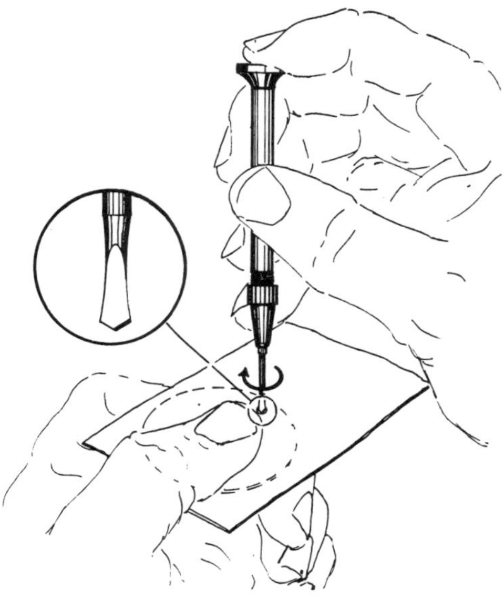

Fig. 4. A drill set into a screwdriver marks the foot position on new dial through holes in cardboard jig.

made by the feet of the sample dial), using a small drill which can be fitted into the nib of a regular screwdriver (as shown in Fig. 4). A countersink tool may also be used.

The marks made by the drill should show slight indentations on the back of the dial, but these indentations should not be too deep or they might show up as bumps on the dial face. These indenta-

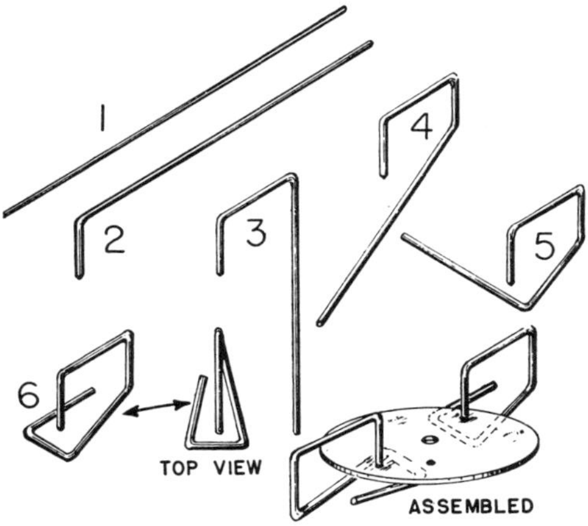

Fig. 5. Steps in making a dial foot clip and method of attaching these to a dial.

tions not only mark the position for the new dial feet but also provide a seating for them, preventing them from shifting to another position before they are soldered into place.

The metal around the new dial feet positions should be scraped clean so that the solder may fuse with the base, insuring greater security for the new feet. The thickness of the new feet is determined by the holes provided for them in the movement. Brass or copper wire may be used for the feet because these metals bend and shape readily, and also take solder without much effort.

The dial feet are prepared as "clips" which may be clamped onto the dial, eliminating any necessity of the feet being held upright and in place by hindering fixtures or jigs. After being soldered, the feet part of the clips are snipped off. The steps in making such dial feet clamps or clips are shown in Fig. 5.

The tip of the clips (which becomes the dial feet) should be slightly pointed so that these fit snugly into the indentation. The clips are then fitted to the dial (as shown in Fig. 5).

Soldering dial feet calls for great skill because during this operation there is danger of "burning" the dial. Too much heat applied in attempting to melt the solder may burn the lacquer on the dial, tarnish the silver—thus discoloring the figures. Thus it is well to use a solder with a low melting point and good fusing qualities. Although there are some bismuth-base alloys containing indium, which melt at the comparatively low melting point of 117° F, they may be too soft and weak; on the other hand, some bismuth-base alloys make good solders, having a higher melting point but still low enough to avoid harming the dial.

SOLDERING DIAL FEET

Below is a table of typical solders and their melting points:

Tin	Lead	Cadmium	Bismuth	Melting Range
5%	95%	——	——	595-555°F
50%	50%	——	——	421-361°F
60%	40%	——	——	376-361°F
19%	31%	——	50%	201°F
13%	27%	10%	50%	158°F

Some good solders of low melting point, such as *Cerroseal-35,* composed of indium and tin, and having a melting range between 240°F and 260°F, have exceptional binding qualities and are safe to use.

Heat applied to a bare metal surface hastens its oxidization, making fusion difficult. A soldering flux is used to prevent oxidization and to permit solder to flow freely and fuse with the dial and foot.

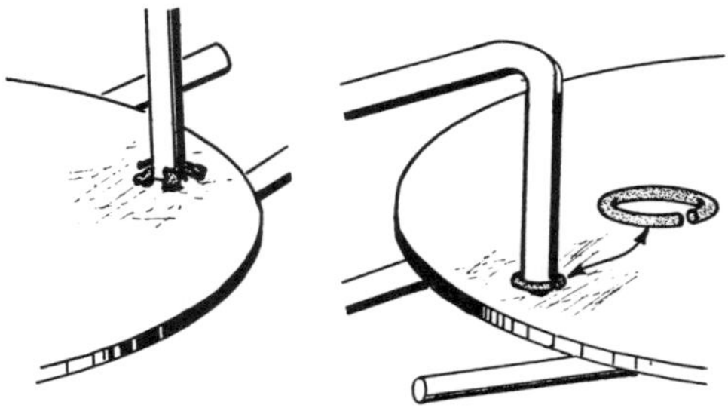

Fig. 6. Solder may be applied to leg either in chip form (left), or as a small ring.

A good soldering solution can be made by dissolving thin strips of zinc in hydrochloric acid. Commercially prepared soldering fluxes (paste or grease form) work equally well, and have the advantage of holding the solder in place until it melts.

After the foot clip is put into place, solder the area around the base of the foot, which has been previously smeared with soldering flux. The melted solder will adhere only to the area covered by the flux. Naturally the amount of solder used should be adequate to bind the foot firmly. Excessive use of solder, however, requires subsequent filing but worse than that may burn the dial. Thus it is well to use a few very small pieces of solder (which melt more quickly) rather than one large piece.

These chips may be placed evenly around the foot, as shown in Fig. 6 (left), so that when they melt this will make a uniform weld around the base of the foot. If solder is available in thin wire form, it may be made into a small ring and placed around the foot as shown in Fig. 6 (right).

MELTING THE SOLDER

A flame is the most satisfactory to melt solder, assuring an even flow around the foot. The flame from an alcohol lamp may be directed by a blowpipe to the spot to be soldered. (A blowpipe is a small tapered "L"-shaped brass tube with a small nozzle at its tip.) The blowpipe is held in the mouth and a stream of air is directed behind the flame which bends to the spot desired. The flame should be soft and broad, rather than pointed, as the softer flame distributes heat evenly. A pointed flame pinpoints heat in

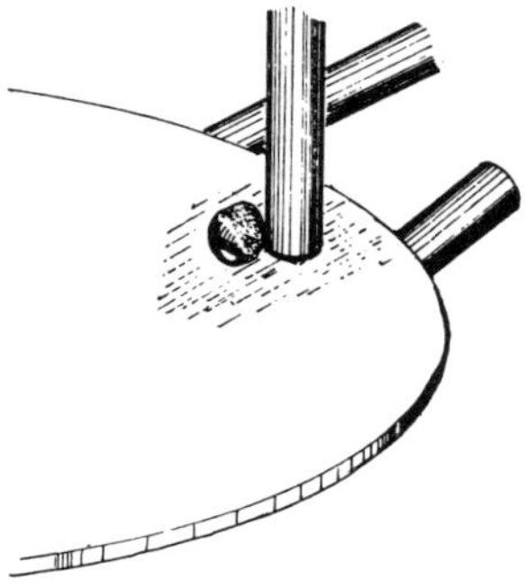

Fig. 7. Solder may roll into a ball if surface is unclean or heat is directed only at solder.

one spot, which may cause the solder to melt before the metal around it has reached a sufficient temperature to effect a weld. This may cause the solder to melt into a ball against the foot, as shown in Fig. 7.

The direction in which the flame is applied to the dial foot should be diametrically opposite the foot, as shown in Fig. 8. This is a safeguard against any tongues of flame accidentally creeping under the dial and burning the dial face. A good weld requires that the metal of the dial near the foot and the leg itself approach the melting temperature of the solder. To prevent the heat from exceeding

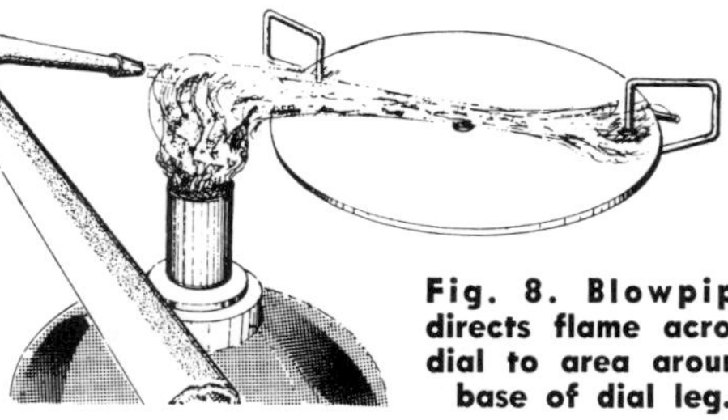

Fig. 8. Blowpipe directs flame across dial to area around base of dial leg.

the melting temperature, the flame is blown at the spot in short spurts, heating the dial gradually so that when the melting point is reached the solder flows and the flame can be stopped at just this instant, without continuing the flame and raising the temperature of the dial.

If two feet must be soldered, they should be soldered one at a time. The second foot can be soldered with less heat because the dial surface is already hot from the heat applied in soldering the first foot. Both feet clips must remain intact until the solder hardens. Using cutting pliers, sever the clip to the desired length. Remove the excess solder, clean the soldering solution or residue paste from the dial, and then try the dial for fit, first in the cardboard tracing and then in the movement itself.

OTHER METHODS OF REPLACING DIAL FEET

Sometimes it is necessary to repair a dial because one of its feet has broken off near its base. This can be replaced by the method described, or by making a dial foot on the lathe and fitting it on

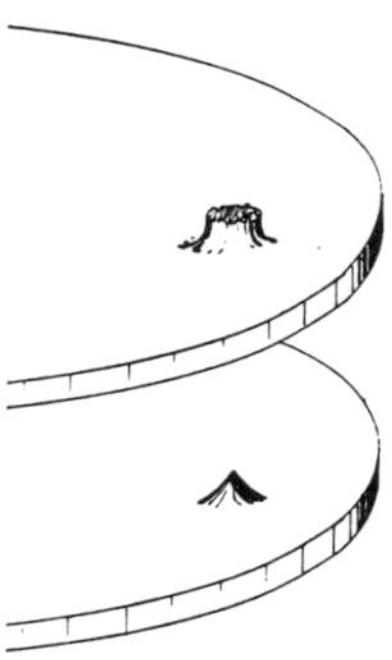

Fig. 9. Preparing broken dial foot stump to receive Self-centering dial foot.

the stump of the broken one. This can be accomplished by filing the stump until a point of it remains, as shown in Fig. 9. Then a dial foot with its base is turned on the lathe, so that the foot appears as in Fig. 10.

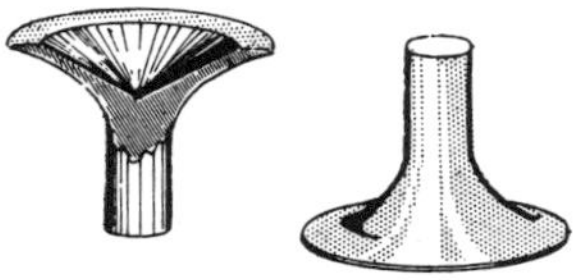

Fig. 10. Special type dial foot which fits over old foot stump and is self-centering.

The countersink on the bottom of the foot fits over the stump, which serves as an accurate guide for placing the foot. Soldering paste and small chips of solder should be placed between the bottom of the foot and the dial. Heat may be applied with a soldering iron pressed on the foot, or by applying an alcohol lamp flame,

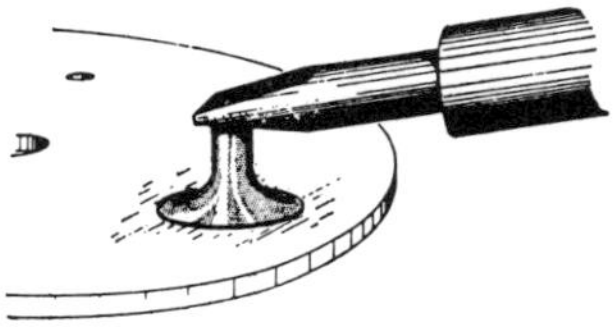

Fig. 11. Heat from soldering iron melts solder under special type dial foot.

using the blowpipe. When the flame is used, clamp the foot into place; when the soldering iron is used, no clamp is necessary because the iron holds the foot in place, as shown in Fig. 11.

The bottom rim of this type of foot must be of smaller diameter than the countersink of the dial in the movement, or the dial will not lie flush with the main plate.

Another type of dial foot fastened in the same way may be used with a regular foot replacement. The foot is prepared in the lathe

Fig. 12. Foot clamp and special type foot which can be made on lathe.

but the lower underpart has a point instead of a countersink. This point fits into the slight countersink cut into the dial as a seating for the foot. A clamp to hold this down is shown in Fig. 12. It consists of a mainspring bent into the shape indicated, with a slight indentation to fit over the top of the foot.

To replace a foot on a porcelain dial, grind away the area around the new foot (which is covered with a thin layer of porcelain), exposing the metal base of the dial to which the dial foot is to be soldered. Use a small grindstone or carborundum pencil for grinding, and do not apply excessive pressure or the dial may crack. The dial should be supported directly under the spot being ground by one of the fingers of the hand holding the dial.

Still another method of securing dial feet is to make legs as in Figures 10-11. These are then placed into the watch and secured by the dial screws so that the tops of these legs are flush with the movement dial plate. Next fill the cavities with epoxy or smear the rims of these legs with Alpha rapid bonding adhesive or similar fast setting bonding material. Next place the dial immediately into position and allow this to remain for as long as necessary. After the solidification period, the dial may be treated as though the legs were an original part of the dial. Make certain that the bonding material doesn't spill over to the movement.

Although excessive heat will not burn porcelain dials as quickly as painted dials, too much heat may crack porcelain. Another precaution: hold the dial foot of the porcelain dial with the cutting pliers about a half-inch above the palm of the hand, and apply pressure gently when cutting the foot. This careful procedure will (1) prevent abrupt severing of the foot and consequent shock which could crack the porcelain, and (2) permit the dial to drop easily into the palm of the hand.

If a dial foot is too long, it may reach too far into the movement and butt against a wheel, upper plate, or bridge, or protrude from the plate, preventing the case from closing and damaging the dial. Too short a foot, on the other hand, will not permit dial screws to grip the feet and the dial will be loose.

REDUCING DIAMETER OF DIAL FOOT

When dial feet are too large to fit into the holes of a movement, the feet can be filed uniformly smaller by skillful use of a short "pillar" file. Stroke with the file (twisting motion) around the feet, removing the metal evenly from them. However, if the original center of the foot is shifted in this operation, the dial will not align or fit properly.

For greater precision use a rose cutter or female broach, a tool shaped like a pinivse but having a screwdriver head for turning easily by hand. The chuck of this tool has three sharp steel jaws cut back slightly to permit them to ease onto the foot, at the same time preserving the tapered base joining the foot to the dial. This tool, shown in Fig. 13, is operated by opening it to fit over the foot and then closed until the jaws just touch the sides of the foot. Repeated twisting and closing the chuck of this tool as the metal is ground off will reduce the foot to the desired diameter and permit it to enter the hole in the movement with a minimum of sideshake.

A new pinvise of good quality also may be used to cut down the

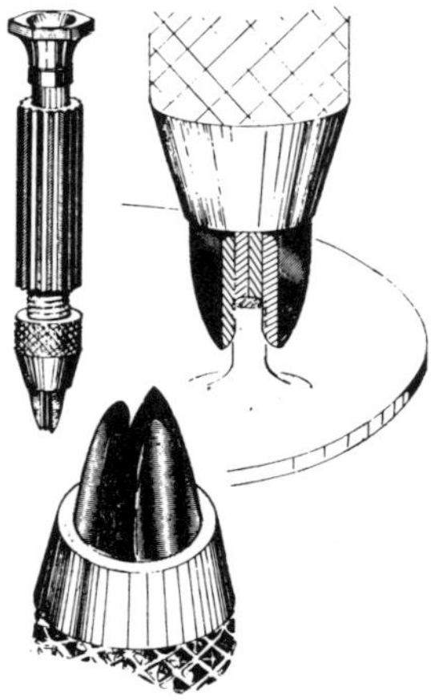

Fig. 13. Special tool used to reduce thickness of dial feet.

size of the foot. Ends of the jaws should be countersunk slightly so that the base of the foot is not weakened. Tools for reducing the size of dial feet, largely Swiss-made, are sold for the same purpose on hour wheel tubes and minute wheel posts.

FITTING A DIAL WITH SCREWS

On very thin watches and on watches having exceedingly thin dials, there is not much room to provide for dial feet. Also the thin dial may bend if dial feet are attached. Secure the dial to the movement by drilling two holes diametrically opposite at points farthest from the center of the dial, usually at the 6 and 12 positions on the dial. Turn the screws into the holes, as shown in Fig. 14, which should be at points on the dial which cannot be seen through the crystal in the bezel. Do the drilling with a pivot drill placed securely into the nib of a screwdriver.

Place the dial over the movement so that the hand holes and crown position are aligned. While holding the dial firmly in posi-

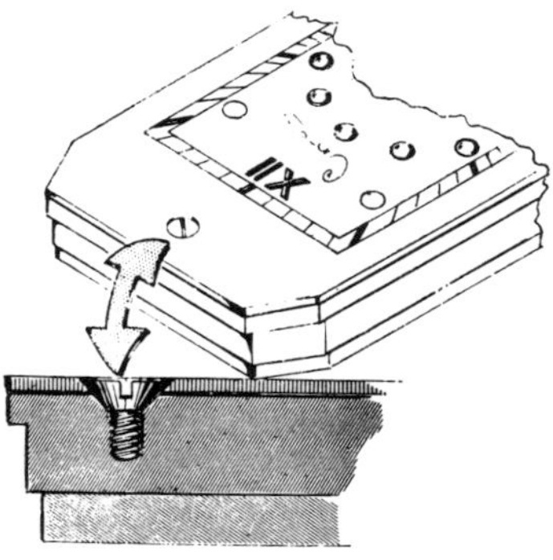

Fig. 14. Top dial screw and method of securing dial to movement.

tion to prevent shifting, spot mark the movement for the holes to be drilled by turning the drill point in the dial holes just drilled. Then drill the holes in the movement. With a countersinking tool, remove all burrs from the holes under the dial and from holes in the movement. Countersink the holes on the face of the dial deep enough for bevel screwheads to seat flush with the face. If left higher than the dial, the screwhead may prevent the case top from closing, or the bezel reflector from touching the dial (as most case tops should). Recommended bevel type screws for attaching the dial to the movement are those usually used to secure setting bridges of small watches. Such screws are usually heavier than ordinary cap jewel screws, and have wider heads for better gripping action. These screws may be turned into the movement, threading the holes at the same time. (It is not necessary to thread the holes in the movement in a separate operation because they are small and the thread pitch will be extremely fine.)

SHAPING A DIAL

Although suitable for a particular watch because its figures, feet, and opening fit, a dial may have too great an overall diameter, preventing the case from closing over it. Fit such a dial by trimming off excess metal with a small metal shears; hold the dial face up in the left hand and the conventional type of shears in the right hand, having the blade of the shears under the dial close to its center. Then the shearing blade on top of the dial will cut along the line desired.

A shears with reverse blades is necessary for left-handed persons performing this operation; if the conventional shears is used in a left-handed manner the cutting blade, being towards the inside of the dial, will throw up an uneven ridge on the face of the dial. Exercise care so that the shears do not interfere with the feet of the dial. For shaping curved dials, special metal shears are available.

When using a file to shape a dial, file the dial from the face side towards the back, holding the file at an angle as shown in Fig. 15, to prevent peeling of lacquer, paint, or silver coating. The file should not be coarse, but a No. 4 or finer grade. Scrape or file off burrs raised on the back of the dial.

ENLARGING DIAL HOLES

Porcelain dials can be shaped by grinding with a fine grindstone. Tilt the dial against the stone, grinding from the face of the dial towards the back. Be careful not to overheat it.

Use a round needle file to enlarge or shift holes in a dial. Many watchmakers insert the file into the holes from the bottom side of the dial; if the file is inserted from the top side, there is danger of scratching the dial by pulling the file back too far on one of the back strokes. Hold the file like a pencil and manipulate it through

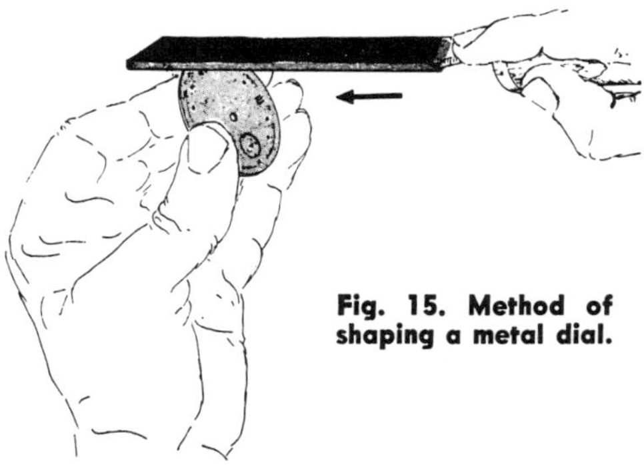

Fig. 15. Method of shaping a metal dial.

the holes in the top side by flexing the fingers while holding the wrist stationary.

A broach or round file may also be used to enlarge the hole in the dial. If a file is used, twist it counterclockwise in the hole. Remove the burrs raised by the file or broach with a countersinking tool (just enough metal to remove the burrs cleanly). Counter-

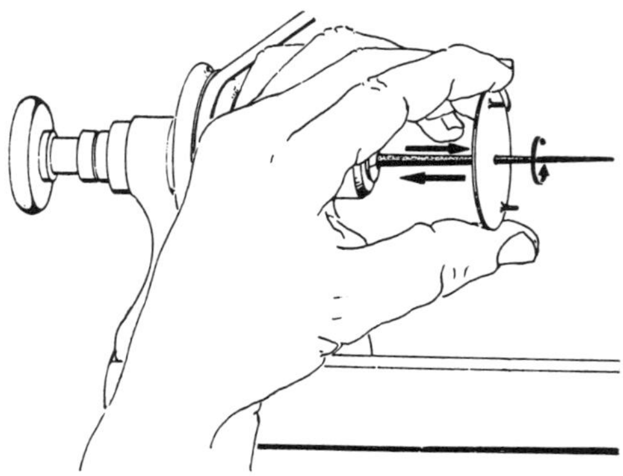

Fig. 16. Enlarging a hole in a poreclain dial by applying abrasive powders to taper-lap.

sinking tools come in sets of three or more and have a small, polished hard steel wheel with sharp edges. The cuts made by the countersink leaves a very high polish. (Fig. 28—Jeweling.)

In porcelain dials, holes can be enlarged or shifted by placing oilstone powder on a tapered iron or copper rod which is turned in the lathe. Add a little oil to the powder for holding the abrasive and for lubrication. Move the dial gently in and out on the taper, as shown in Fig. 16, being careful that the taper doesn't "grab" the dial, which might crack the porcelain.

REPAIRING PORCELAIN DIALS

Porcelain dials, used mostly on older watches, are becoming scarcer and more difficult to replace. When such a dial becomes chipped or cracked, it is possible to repair it by filling in the cracks or chipped area with an enamel that comes closest to matching the porcelain on the dial. This enamel is sold by watch supply houses and is available in two types, one a liquid applied with a brush and the other which resembles wax. The wax-type enamel in various colors is sold in little tin boxes marked *Kalte Emaille or Email Froid,* or "cold enamel," its English equivalent.

Before applying the enamel, clean the dial as thoroughly as possible, leaving no lint or fingerprints near the spots to be repaired. Place over the crack or chipped area a piece of enamel, free of dust and clean, large enough when melted to fill the area. Hold the dial over an alcohol lamp flame so that when the dial becomes warm, the quick melting enamel flows and completely fills the crack in the dial, as shown in Fig. 17.

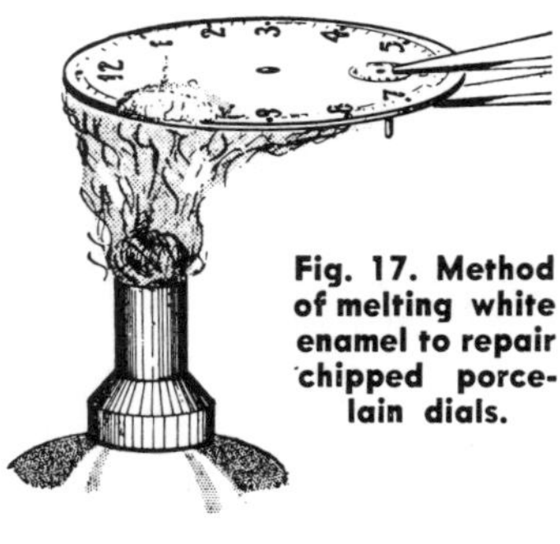

Fig. 17. Method of melting white enamel to repair chipped porcelain dials.

Remove the dial from the flame as soon as the enamel melts, because the enamel may become thin and will spread if the dial is held over the flame too long. After the dial is taken from the flame, the enamel cools quickly, forming a slight mound over the spot where it melted. Cut away the excess enamel with a very sharp, smooth-edge razor—skimming the razor over the dial—leaving the enamel level with the surface of the dial.

If the chipped or cracked area is over a numeral or the minute track, paint in the numeral or minute track with India ink or liquid enamel, using a small brush down to a fine point. Do not use a pen for it scratches soft enamel, and tends to spread the liquid coating. Spray or paint the surface of the filled area with a thin layer of clear lacquer to preserve the numerals and keep the new enamel clean. While such a job may not be perfect and slight flaws may be detected with a loupe, it is a great improvement over an unsightly cracked dial.

Else, several thin coats of "Appliance Touch-Up" paint, available in hardware stores is good, afterwards the excess is razored off and polished as you would an acrylic plastic.

SHIFTING DIALS

While it is sometimes necessary to file a hole in the dial to shift it to one side to avoid having the hole touch the hour wheel, enlarge the hole only after it is certain that other methods will not do. Filing a hole is necessary only when the dial is off center, or the center and second hand holes must be shifted.

If a dial hole crowds the hour wheel on one side, the dial feet are either bent or incorrectly placed. (This may be verified by checking the crown position and the second hand hole.) If the holes for the hour wheel and the second hand are both off-center in the same direction, straighten the feet of the dial, or fit the dial to the movement, tapping the edge of the dial in the direction that centers it. By this procedure the dial feet will be bent from their former position as shown by the dotted lines in Fig. 18. The tapping is done best with a fibre or horn hammer. The balance of course should be removed from the movement before this operation.

Some older watchmakers, and those specializing in watch casing and dial fitting, have special pliers to shift the dial feet, as shown in Fig. 19. Those specializing in casing many movements of one kind use a metal block, the shape of the movement, into which two accurately placed holes are drilled for the dial feet. The dial is placed on the block with the dial feet in these holes, and the dial is tapped until it is centered, relieving the watch movement of shocks accompanying this correction.

CLEANING DIALS

Almost all dials are made with a brass base upon which a thin layer of silver is deposited. The figures are printed upon the silver deposit with a "soft enamel" paint. The printed surface is sprayed with clear lacquer to protect it from dirt and oxidation. If the lacquer peels or becomes porous after aging, vapors, acids, moisture, dirt, and other matter may discolor the dial.

Fig. 18. Bending a dial foot by tapping the dial with a horn-head hammer.

Many chemicals used to remove dirt, grease, and oxidation from plates and wheels cannot be used on watch dials because the chemicals may damage the painted numerals. Watch cleaning solutions composed of ammonia and soap, and a water base, may be used to remove dirt or tarnish if the solutions do not contain acetone, alcohol, or benzol (which may attack the dial figures).

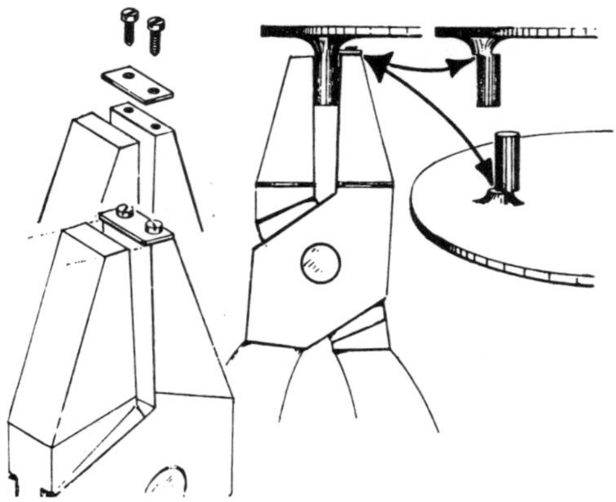

Fig. 19. Method of converting pliers, which may be used to offset dial feet.

If this chemical content of the solution is slight, the dial may be dipped into the solution and swished a few times; but dry by blotting gently (do not wipe). Do not use alcohol; it dissolves paints and lacquers.

Cyanide solutions remove tarnish from a dial although their use is discouraged for anyone but those thoroughly experienced with them, as they are deadly poison.

Do not wipe dials with radium figures after cleaning, as the radium compound is softened by cleaning liquids and may be smeared. Blot such dials with a clean chamois or clean linen cloth.

Solid silver dials with hard enamel figures may be "refinished" by placing the dial in a copper pan filled with a solution of 10% sulphuric acid and water, as shown in Fig. 20. This solution is heated with the dial submerged in it until the dial becomes snow-white, after which it is removed, dried, and lacquered. Some watchmakers prefer to heat the dial and then drop it into this solution; however, the immediate quenching may crack the enamel figures. Heating the dial up to about 700°F will not harm the hard-baked enamel figures.

In preparing the sulphuric-acid-and-water solution, be certain to place the water in the pan first and then add the acid. If the reverse is attempted, a great heat will be generated with accompanying hazards that could be exceedingly harmful to the watchmaker. In any case, no one should attempt to work with chemicals until he has become thoroughly familiar with their working properties and dangers.

Fig. 20. Solid silver, hard-enamel dials may be renewed by boiling in weak sulphuric acid solution.

Some watchmakers achieve the same finish on silver dials by using a cyanide solution. The action is much the same, a slight layer of the silver being removed chemically, exposing a bright fresh layer of silver. Lacquering the dial protects the new surface from tarnish.

BOOK III

CHAPTER V

HOLLOW CENTER AND CANNON PINIONS

ADJUSTMENTS AND REPAIR

THERE are two types of cannon pinion tension which permit the going train's motion to be transferred to the dial train. In this way, the turning center pinion carries the cannon pinion and will permit the hands to be set by the setting mechanism with the desire of the wearer.

The most popular system is shown in Fig. 1. The center pinion and the arbor are one solid piece, reaching up through the dial

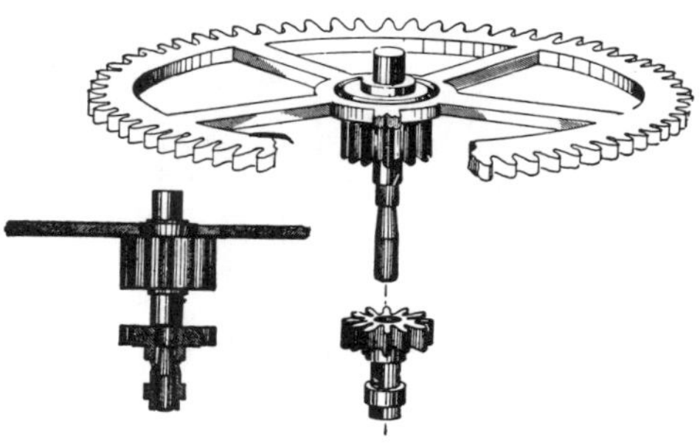

Fig. 1. The Snap-on type of cannon pinion which fits over the center pinion post, snapping into place when its dent fits into the groove in the one-piece pinion and post.

plate. This cannon pinion, known as the snap-on-type, slipping clutch-tight over it, has sufficient grip on the arbor to be carried along by the revolving center post. In turn, the cannon pinion turns the minute wheel, the minute wheel pinion engaging the hour wheel. However when the hands have to be set to a new position, the turning of the setting mechanism by the crown overcomes the grip of the cannon pinion on its post and forces the cannon pinion to slip around the center post.

Should the tension grip of the cannon pinion on the centerpost weaken to the point where the drag of the dial train wheels on the cannon pinion becomes greater than its grip on the post, the center post would then continue to turn without carrying along the cannon pinion and the rest of the dial train.

To increase this tension, the thin springy neck of the pinion is indented so that the inside diameter of the tubular hole is reduced. This enables it again to grip the center post with the appropriate tension.

Another system used to obtain cannon pinion tension is that which is used with the hollow center pinions (Fig. 2). It is em-

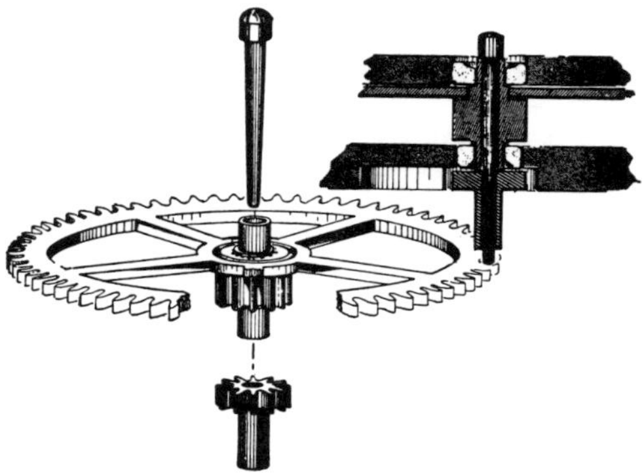

Fig. 2. The Hollow Pinion type of Cannon Pinion showing, top, the tapered center-pin, the wheel with its attached hollow pinion and the straight-walled cannon pinion. Its assembly between the plates is shown in the upper right.

ployed in stem-wind-set older watches and also in watches and clocks in which the hands are set directly from a knob fastened over the rear of the center-post which extends through the back of the case.

The tension in this system is effected in a much different manner from the snap-on type cannon pinion. The center pinion with its upper and lower pivots is made hollow by having a hole drilled through it. A slightly tapered steel pin fits through this hollow pinion and above the dial plate. A thick walled cannon pinion is forced over this tapered pin so that it is gripped very tightly.

Usually, the tension is located between the walls of the hollow pinion and the tapered pin-post. This tension is sufficient to enable the turning center pinion to carry along this pin and its mounted cannon pinion as well as the minute and hour wheels. When the setting mechanism activates the dial train, the cannon pinion, fast on the tapered pin, carries this pin along, rubbing against the inside walls of the hollow center pinion. Constant setting of the hands will eventually cause the tapered pin and the inside wall of the center pinion to wear away. Since this center pin is tapered, it also must not be permitted to slip down inside the center pinion as to do so would cause it to become loose and though the movement would go, the hands would remain motionless.

By fitting the cannon pinion over this pin so that it goes fully down and contacts the top of the hollow pivot while the pin-head hugs the top of the opposite pivot, the tapered pin will be prevented from working outward and loose. The dial train then cannot be carried along by the revolving center-pinion. In such a case, the pin must be removed and replaced with a better fitting one or its thickness sufficiently increased to again fill up the hollow space inside the center pinion.

Even though the pin might be loose inside the center-pinion, it is still tightly wedged in the cannon pinion. Therefore removal of this type of cannon pinion should not be attempted in the same manner as the snap-on type as this might result in "springing" the main plate. This type of cannon pinion assembly is dismantled by pushing out the tapered pin from the cannon pinion and the center pinion, sometimes without taking the center wheel out of the movement.

To remove the tapered pin, the staking set is set up so that a hollow stump is lined up opposite a flat-faced solid punch. The

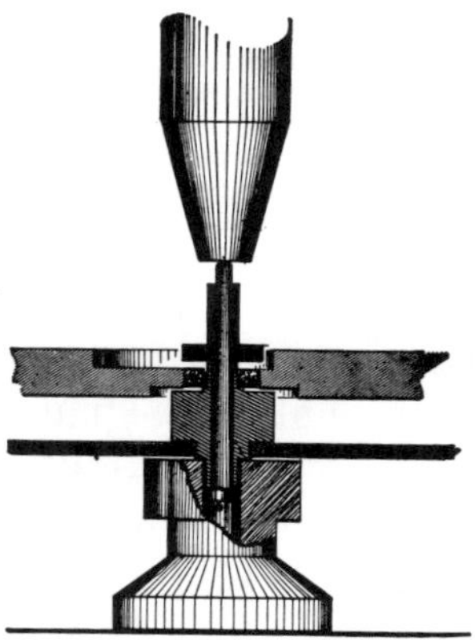

Fig. 3. Ejecting the tapered pin from the cannon pinion and hollow center pinion is done with pinion resting on stump in staking set.

diameter of the hole in the stump should be as small as possible and yet allow the head (back) of the pin to enter it without binding. This is shown in Fig. 3.

Then, while the movement in this position is steadied by the fingers, the punch is tapped lightly. This will eject the pin from the cannon pinion and loosen it in the center pinion enough to be removed from the rear of the movement with a pair of tweezers.

To broaden this pin so that it can again grip the hollow of the center pinion, a flat-faced solid stump is placed in the staking set directly in line with a solid, flat-faced punch of the same diameter as the stump. The pin is placed between these, and a tap of the punch with a hammer will spread the pin enough to add the thickness required. The pin must be placed on the stump as close to its head as possible so that the renewed pin's thickness will be as close to the lower pivot as possible. The punch and stump must be of the same diameter and placed directly over one another or else the pin will become bent. This operation is shown in Fig. 4.

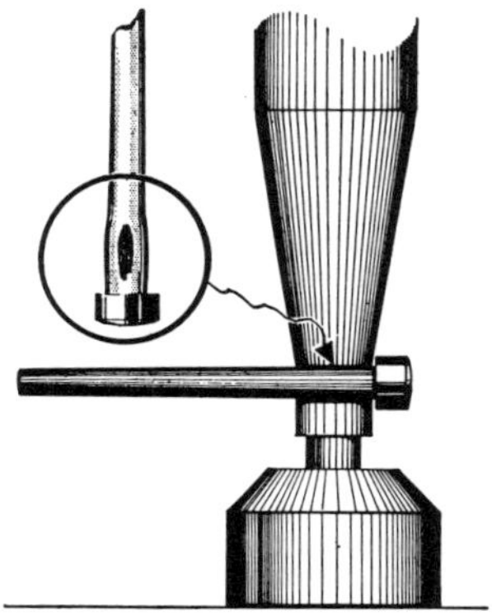

Fig. 4. Increasing the thickness of the tapered pin is done by swedging between solid stump and punch in staking set.

Although the pin is flattened just a little, the metal displaced will broaden it enough to enable it to grip the inside of the center pinion with renewed and adequate tension.

Sometimes it is necessary to replace such a pin because of breakage or because the old pin is beyond adjustment or repair. Square head center pins with a button on them break more often than those with plain round heads. Center pins could be purchased at one time in a boxed assortment under the name of "Center Pins" or "Center-Squares", but no current catalogue lists them. In order to replace one today, it has to be made. This is not difficult or lengthy if the steps to make it are carefully followed.

The first step in making the pin is to be sure that the thickness of (steel) drill rod used should be at least as thick as the head. If the head of the pin must be square, the diameter of the stock used must be at least the same as the diagonal of the square. To obtain the dimensions of the round stock needed to make a square of given size, multiply one side of the square by 1 2/5 or more precisely, by 1.4142. For example, if a square of 1.00 mm. is needed, the round stock must gauge 1.41 mm. (See Fig. 5.)

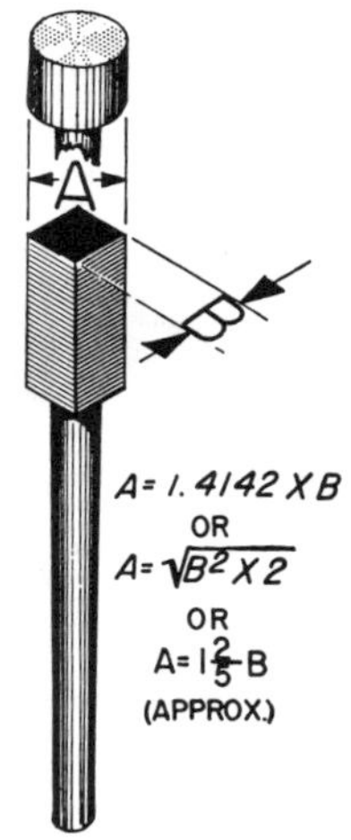

Fig. 5. Formulas for obtaining thickness of round wire needed for square of given size.

If the broken center-pin served satisfactorily before it broke, it may be used as a sample and the measurements, shown in Fig. 6, taken for the new pin for head size (A), length(D), taper thickness at its base (B), and at its tip (C), and the height of the head (E).

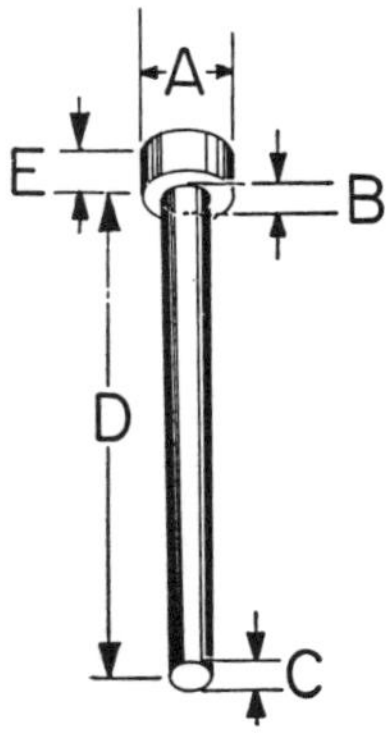

Fig. 6. Gauges for a new center (taper), pin must include head thickness: A, taper thickness at base; B, thickness at tip C; length of taper D; and height of the head E.

Should a replacement be necessary because the taper is incorrect, faulty performance or the sample is lost, then the measurements for the new pin must be determined directly from the center wheel pinion and the cannon pinion; the parts into which it must fit. The metal is placed into an appropriately sized chuck for turning in the lathe. To begin the job logically, the first part of the taper must be turned so that it will fit into the back pivot of the center wheel pinion. The metal must be turned to a fine, gradual taper so that it will become tight when the back end of it is pushed through the rear of the pinion as shown in Fig. 7.

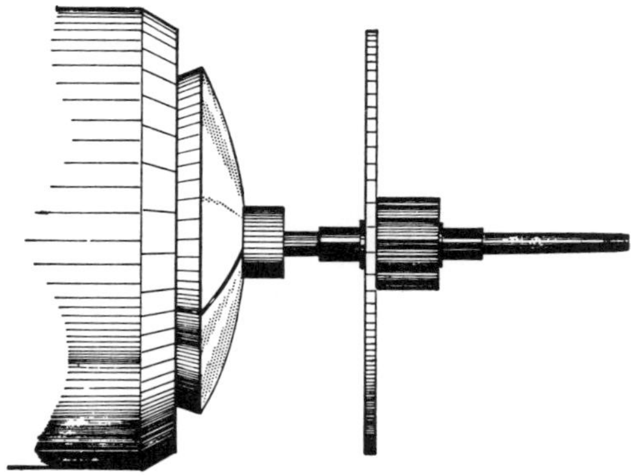

Fig. 7. When taper is turned accurately in lathe, the center wheel and pinion should reach down to position as shown before it meets any resistance or pressure to drive it on further.

Turning a comparatively long thin taper on the lathe calls for skill and patience. The taper must be cut, with the graver cutting from the taper-tip towards the lathe-headstock in order to avoid bending or breaking it. It also must be cut very gradually or it will be faulty for two reasons. The first is that it will bind suddenly without any tolerance for tension in the hollow pinion. It will also spread out the lower center pivot and crack the lower center-hole jewel. Also, a larger (angle of) taper loosens easily. The second point of error

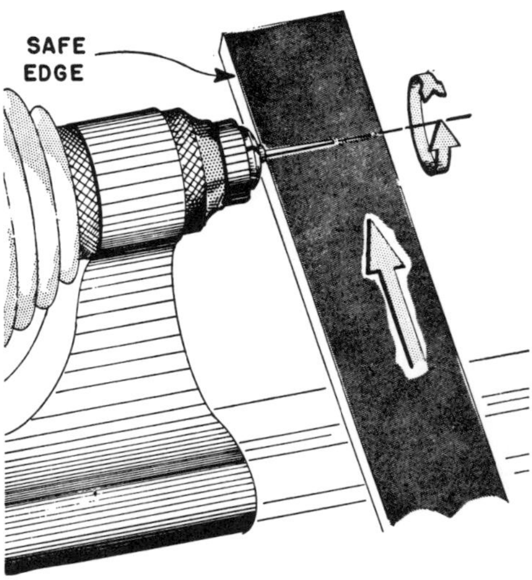

Fig. 8. When taper is formed with file, the lathe motion is reversed, turning in direction shown by the arrow. File is moved lightly in direction indicated on it.

is up in the cannon pinion. If the taper is too great, it will be too thick for the lower center pivot and too thin for the cannon pinion where it will be very loose.

If after turning the taper, it is tight in the center pinion and too loose in the cannon pinion, it is an indication that the taper angle has not been made gradual enough. A new taper can be cut into the stock held in the lathe chuck.

Some watchmakers experience difficulty cutting a taper with the graver and prefer to finish the taper, using a broad, flat, fine-cut file with a safe edge. The file is held under the stock while the lathe rotation is reversed in order for the turning stock to face the cutting edges of the file as shown in Fig. 8.

When the taper is made by this method, the file used must not be coarse but rather fine so that the thin pin being formed does not become caught between its cutting ridges. The safe edge should be presented towards the lathe headstock so that the taper is not made longer and the head shorter and prevent filing into the front of the lathe chuck. Excessive pressure must be avoided in cutting the tapered surface with a file. To do so may break the pin or cause

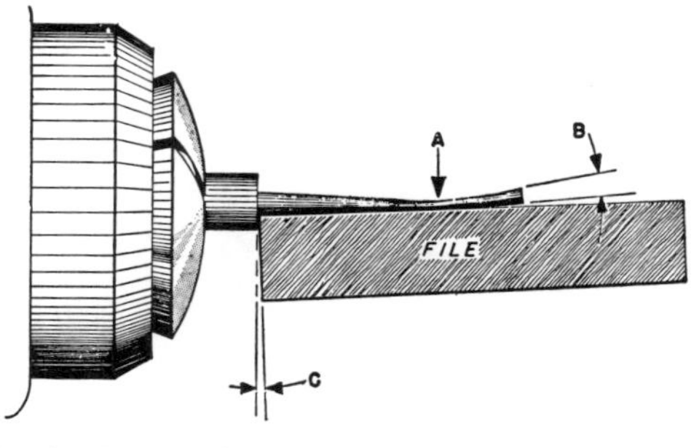

Fig. 9. Too much pressure should not be exerted on file or else false taper will result with A being thinnest part.

the pin to assume a false taper angle with its thinnest part about a third back of the pin-end. This fault is shown in Fig. 9. Because the file is held underneath the work, the filings will fall on the file, clogging it and scratching the work. Therefore these should be cleared as often as necessary.

The taper whether cut by graver or with a file, should be left a little oversize so that it can be smoothly finished to exact dimensions with an oilstone slip or an iron lap charged with oilstone powder. The base of the taper (where it meets the head or square) must have a sharp square shoulder. If the base has a filet, the pin

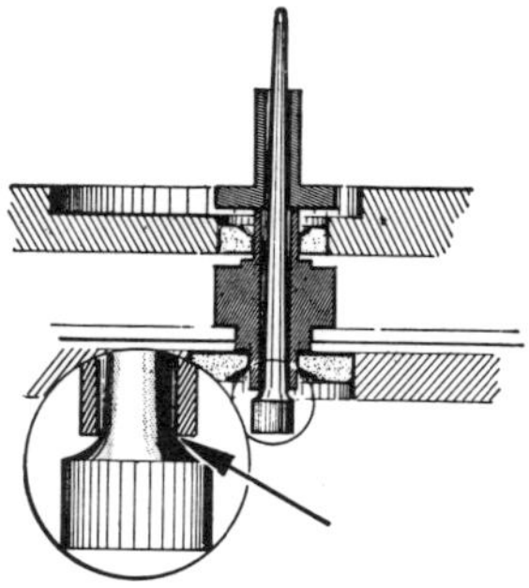

Fig. 10. Center taper with filet at base results in false fit and does not permit this pin to enter the pinion fully.

will not reach up to the pivot of the center pinion and may result in a false tension and soon work itself loose (Fig. 10).

If such a pin is forced in tightly, its curved base will spread the hollow pivot, which in turn will cause it to crack the center jewel as

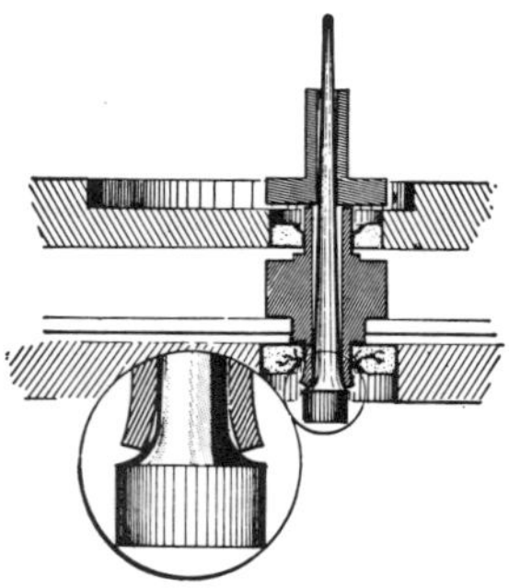

Fig. 11. If filet base is forced into the hollow pinion, it will spread out the pivot and cause jewel breakage or bearing damage.

shown in Fig. 11. If this pivot is housed in an unjeweled bearing, it will make it almost impossible to remove the center wheel from the train bridge.

If the pin's taper is finished to a satisfactory angle, it should be loose when inserted in the center pinion up to three-quarters of its length. When the last quarter of this taper is pushed into the center pinion, it should have some resistance and yet permit twisting within the pinion without requiring too much force.

The cannon pinion is placed on the taper for a trial fit before the stock in the lathe is parted from behind the head of the taper. It should fit down on the fully inserted taper (in the center-pinion) about 2/3 of the cannon pinion's length. This indicates that it would become very tight when forced down so that its bottom touches the top of the center pinion pivot. In this position, the cannon pinion should grip the center pinion considerably tighter than the tension that the taper pin's grip has on the inside of the center pinion. If this tightness is achieved, it will enable the cannon pinion to carry its tapered center pin along when the setting mechanism is operated to set the hands.

The length of the center pin should be sufficient to extend above the cannon pinion when it is tight so that it will provide an extension on which the minute hand is generally mounted. If any length of taper extends beyond the cannon pinion in excess of the thickness of the minute hand socket, it should be stoned off to a nice finish.

A study of this system will reveal that the hole through the cannon pinion should be slightly smaller than that through the center pinion. If not, the cannon pinion will be loose regardless of any angle of taper produced. In order to bring this relationship to the desired proportions, the cannon pinion will have to be changed for one with

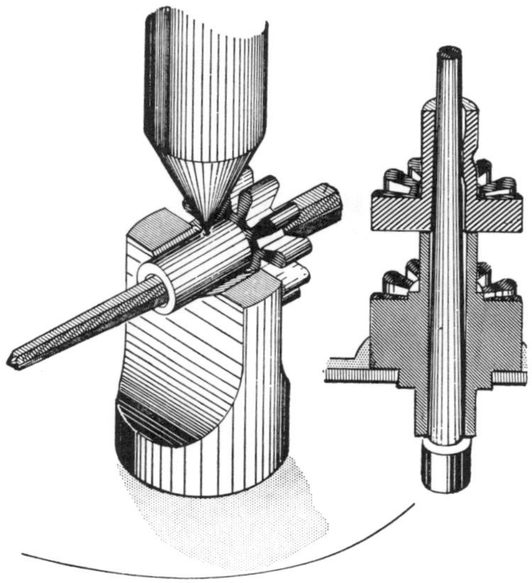

Fig. 12. Tighten cannon pinion hole by indenting it with prick punch (effect shown in assembly at right).

a smaller hole or its hole will have to be made smaller. This can be accomplished in effect, by placing a dent in the lower side of the cannon pinion with a prick-punch in the same manner as in the springneck type cannon pinion. This is shown in Fig. 12. If a broad type of punch is used, it will reduce the effective hole size but it will spread or broaden the outside diameter of the cannon pinion, preventing the hour wheel from slipping over it.

The center pin is afterwards supplied with a head which should be no thicker than the pivot against which it rests (Fig. 13). It is

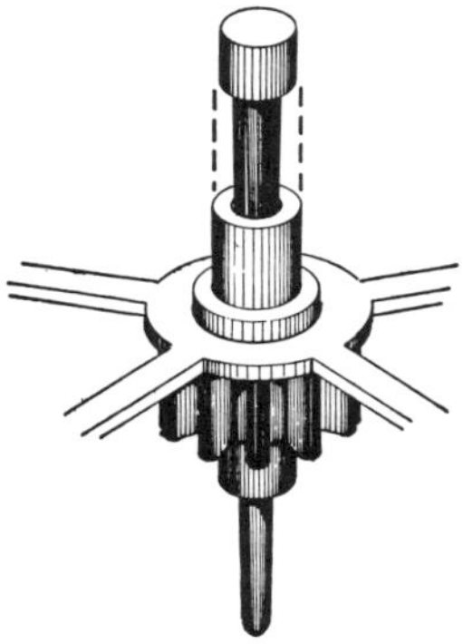

Fig. 13. Taper head should be slightly smaller than thickness of pivot against which it rests.

then parted from the stock and hardened and tempered to a blue color, after which its head is polished or burnished. The tip of the taper should also be polished as well since it is seen protruding slightly through the socket of the minute hand.

CENTER-SQUARES

If a center pin is to be made for an old watch or for a modern table type movement which is set with a key from the back, the head of such a pin must be square. When the size of the square is left to the watchmakers judgment, it is advisable not to have the thickness of the square exceed twice the thickness of the taper's base. Too thick a square places a strain on the taper and increases its susceptibility to breakage. The length of the square should be no longer than necessary as the longer squares on these tapers break more frequently than do the shorter ones.

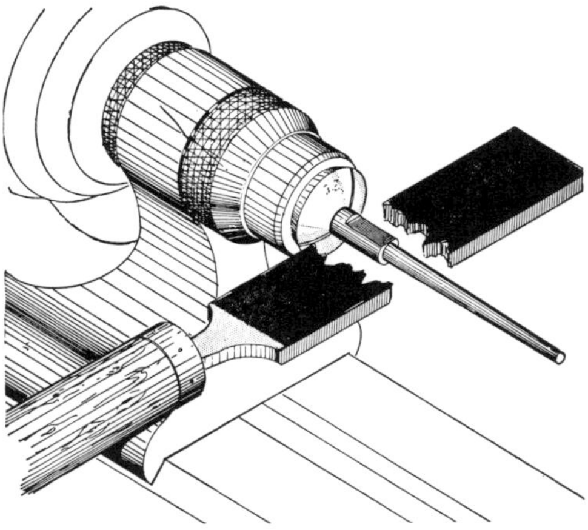

Fig. 14. File square on taper, holding in lathe chuck to avoid bending or breaking.

To make the square, the taper is first formed by the methods described earlier. A section of the metal rod in the chuck behind the taper is advanced to become the square. Some watchmakers prefer to cut off this extra length with the taper and then holding the tapered section in a chuck while the square is filed. However, this places a strain on the thin taper in the chuck. To avoid this, the square should be filed after the taper has been finished while it is still in the chuck as shown in Fig. 14. In this manner, the square may be filed without strain on the taper. The square can be filed with the headstock locking pin set in the quarter holes of the spindle's index plate set into the head of the pulley. (For detailed methods of filing a square, see pages 82 to 87 in the author's "Watch Repairer's Manual".)

The square is finished with a slight taper to allow the setting button or key to slip on without undue force. It should then be hardened and tempered in the same manner as in a regular center pin. To prevent the bottom of the square from scratching the watch plate, a small, thin washer is placed between the square and the

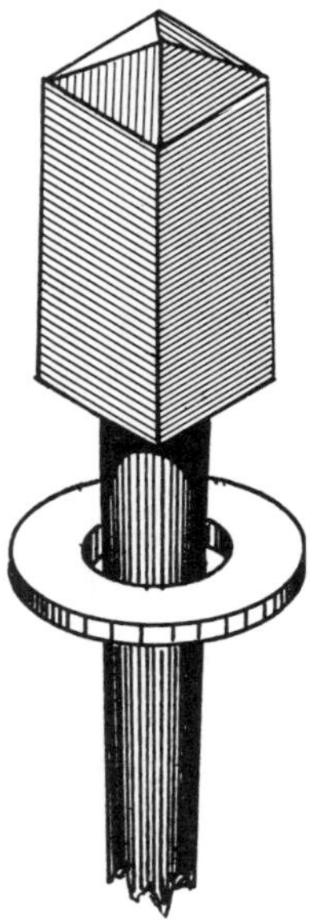

Fig. 15. To prevent scoring plate, separate square from pivot with thin washer.

hollow pivot-end (Fig. 15). This washer must not be too thick or it will not permit the base of the taper to enter the pinion, causing it to become loose. Instead of a washer, it is also possible to start the square further back on the head of the taper and thus provide

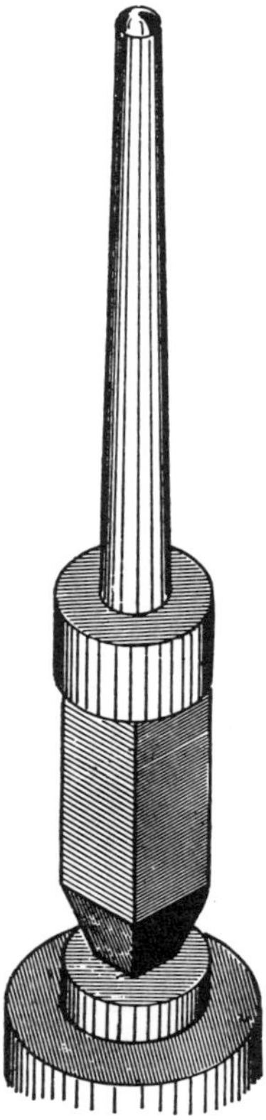

Fig. 16. Washer is unnecessary if square is filed back of taper sufficiently to leave a flat round surface.

a flat-round surface to the end of the hollow pivot as shown in Fig. 16.

CLOCK CANNON PINIONS AND THEIR TENSION

The method of obtaining the tension between cannon pinion and center pinion in most clock movements is shown in Fig. 17. This is best understood if it is explained from the view of logical assembling of this unit.

A thin cylindrical rod of wire acts as the center arbor. The cannon pinion is a brass pinion which is driven onto the center arbor frozen-tight. Any movement of this pinion then carries the center arbor with it, and any movement of the center arbor will also carry the cannon pinion along with it as well. The cannon pinion in this case is but a small wheel without any tube such as is used in watch movements.

The center arbor goes through the lower (dial side) plate. Then a thick brass retaining collet or washer is driven onto the arbor frozen-tight to a point which will allow a certain amount of endshake between it, the plate and the cannon-pinion wheel. Over this washer is slipped a spiral tension-thrust spring, a loose thin steel washer being placed over the arbor and resting on the spiral spring.

The center pinion and center wheel are then placed over the arbor and this rests on the thin steel washer which serves to separate the spring from the leaves of the center pinion. Then another thick brass retaining washer is driven over the center-arbor until it not only touches the back of the center wheel but presses it down a little, squeezing in the spiral spring. This washer regulates the endshake and provides the amount of clutch tension needed. The washer is driven down to its correct relative position frozen-tight.

The upper plate is then assembled after which the hand setting button is slipped over flats in the end of the center arbor.

The friction tension then is able to work in the following manner. The spiral spring pushes against the center pinion on one side and

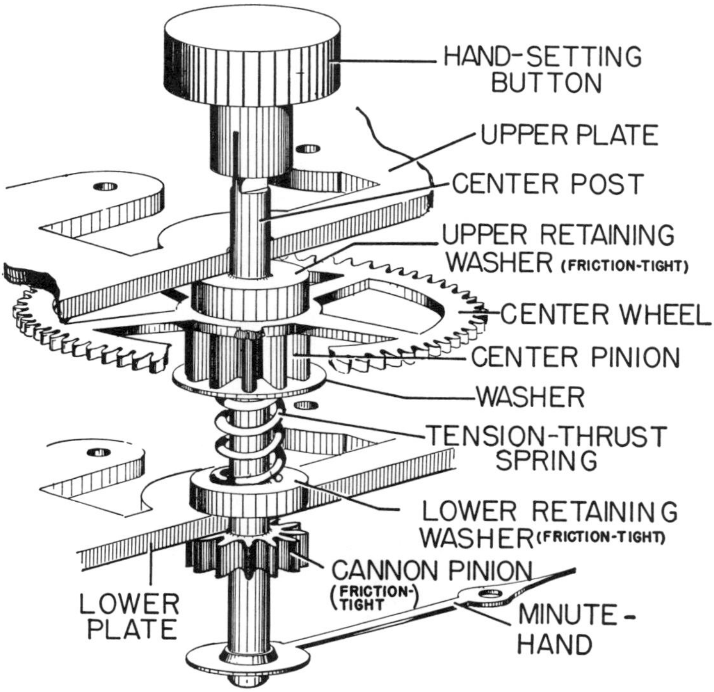

Fig. 17. Center wheel, arbor, and cannon pinion assembly on clock and desk-watch movements, showing cannon pinion friction.

the lower retaining washer on the other side. The center wheel is backed up by the upper retaining washer. When the center pinion turns, the entire unit turns with it because of the spring pushing up against the center wheel pinion. The center wheel and pinion, pressed up against the retaining washer carries that along. Since the retaining washers are frozen tight on the arbor, anything attached to the arbor in this manner will turn with it. This includes the cannon pinion wheel.

However, when the hand setting button is turned by the fingers, the tension of these pressing springs is overcome by the force of the fingers turning the button and the arbor skids inside the center pinion and washer rubbing over this spring. The minute hand, mounted on the tip of the center arbor will turn with it regardless which force is applied to the center arbor. The fast cannon pinion wheel turns the dial train which activates the hour wheel and the alarm mechanism.

Should this tension be lost or weakened, pressing the spring together by driving the washers closer together will tighten up on the tension. However, this increases the endshake and the best solution is to remove the upper retaining washer, obtain a higher spring and re-assemble the unit as shown in Fig. 17.

Some older watches obtain the tension by using a spring washer between the head of the taper pin and the pivot-end. The tension is renewed by reshaping the washer and driving the cannon pinion and taper pin head together, which depresses the spring-washer.

A system employed by some of the finer watches which use the hollow center pinion method is shown in Fig. 18. Here the cannon pinion and taper pin are made from one piece. This is inserted through the dial side of the movement and the friction takes place

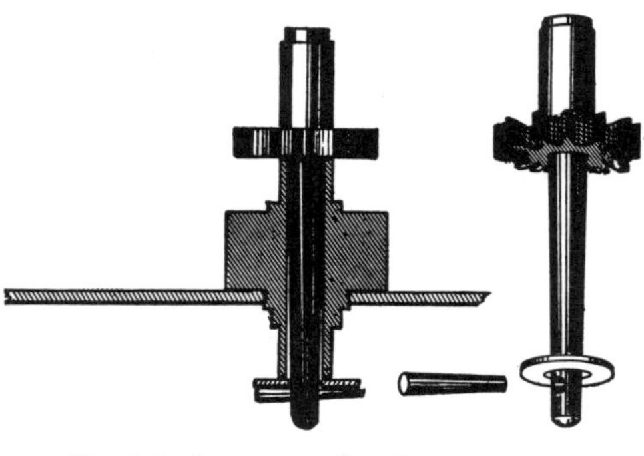

Fig. 18. Some watches have cannon pinion and tapered center pin in one.

closer to the dial side hollow pivot of the center pinion. To keep this in place, a thin short taper pin is inserted through a hole near the end of the taper and is separated from the pivot by a thin small washer. Renewing the friction on such an arrangement is done in the same manner as shown in Fig. 4 except that the slight bulge is placed as close to the cannon pinion as possible.

To assemble a cannon pinion on a hollow center pinion taper pin assembly, the minute wheel is first removed to avoid crushing its teeth when the cannon pinion is pushed down. Then the staking set is set up with a tapered-side solid concave stump, centered against a flat-faced, hollow punch whose hole fits closely over the cannon pinion tube as shown in Fig. 19. The movement is then placed over

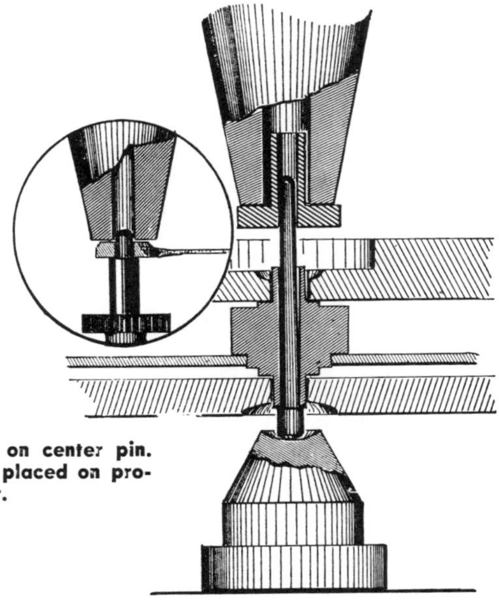

Fig. 19. Replacing cannon pinion on center pin. Inset shows minute hand properly placed on protruding center taper.

the stump so that the (bottom) head or square of the center pin rests in the concave of the stump. The punch is then placed over the cannon pinion and a few light taps on the punch should be sufficient to bring the cannon pinion down tight and snug against the (dial side) hollow pivot and the taper-head against the opposite pivot-end. If the hammer blows during this operation are too strong, there is danger of mushrooming both pivot-ends.

Replacing the minute hand is done in the same manner as replacing a cannon pinion but with a hollow punch whose hole size is just a little larger than the thickness of the protruding center (taper) pin. This operation is shown in the inset of Fig. 19.

Book III

Chapter VI

REPLACING WHEEL TEETH

Since watchmakers are called upon to do a great amount of work on clocks, some instruction on repairs other than normal interchangeable replacements should be provided. Suffering less abuse, clocks tend to remain in use longer than watches. Consequently, there are literally thousands of different models no longer made which need servicing. Parts are not available for them, however, and the watchmaker must know how to do the repair effectively and economically.

Because of the extra strong mainsprings and weights which drive these clocks, wheel teeth tend to break when subjected to sudden spring or weight slippage or power backsurge. Since it is both costly and time-consuming to have a new wheel cut, competent repairmen know how to replace wheel teeth. Usually, the wheel that suffers most is closest to the power sources—either the mainspring barrel or the main wheel. These are often the largest wheels in the clock. With a little knowledge, fitting new teeth in them should be easy.

REPAIRING MINOR DAMAGE

If damage is confined to only a few teeth, a dovetail is cut out of the rim and a sheet of brass is cut to fit this dovetail. The teeth are then cut and filed to shape.

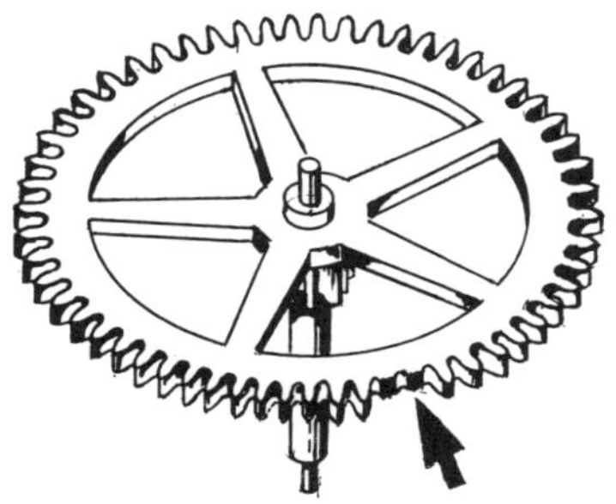

FIGURE 1
A new piece of brass will have to be dovetailed into this wheel and new teeth cut to replace the broken ones.

Figure 1 shows a wheel with two teeth broken, indicated by the arrow. First a dovetail slot is cut into the rim of the wheel. It is made so that only enough metal is removed from the wheel to provide a secure fit for the plug. Notice the well proportioned dovetail slot in Figure 2. The width of this slot, *A*, is barely wider than the space occupied by the two missing

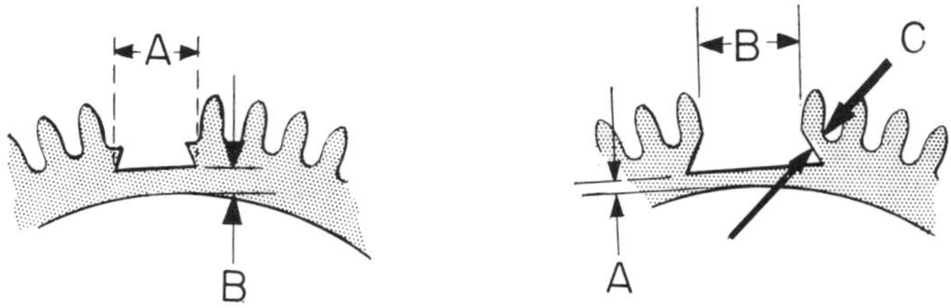

FIGURES 2 AND 3
The dovetail should be well proportioned, as in space A. Space B undercuts the adjacent teeth, weakening them.

teeth. The depth of the cut, *B*, still leaves sufficient metal to the rim so as not to weaken it.

Making the dovetail slot as wide and as deep as in Figure 3 introduces two faults. Here, too little distance remains between the cut and the rim at *A*. This may distort the concentricity of the wheel rim as well as the wheel itself when a new plug is fitted. The dovetail slot was also cut so wide at *B* that it undercuts the adjacent teeth, leaving a weakened area at the spot indicated between the heavy arrows at *C*. Should a plug be fitted to this wheel with such a cut, the two adjacent teeth will be bent, as shown by the arrow in Figure 4.

Figure 5 indicates the correct proportions to follow in cutting this important slot. Using a fine saw, begin cutting the dovetail at the base of the broken tooth. Cut on the diagonal so that the slot's depth is only half-way down the width of the wheel rim. The slant of the dovetail's angle should not intrude on the adjacent tooth's radial line, in Figure 5. The saw cutting should be modest, since it is best to finish the shape of the dovetail with a file.

FIGURE 4
When the plug is inserted, neighboring teeth bend easily from the strain of poorly proportioned dovetail.

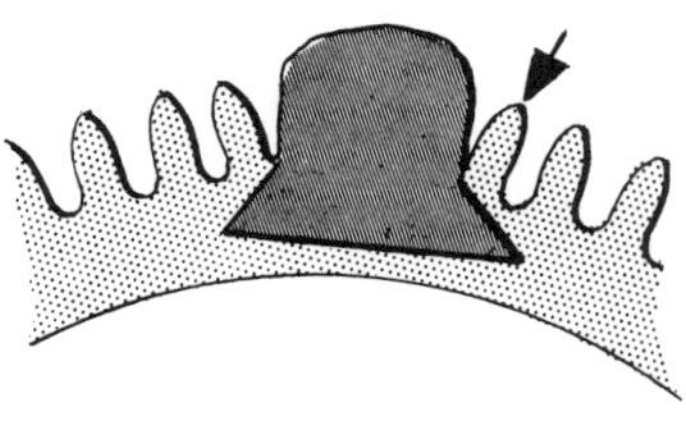

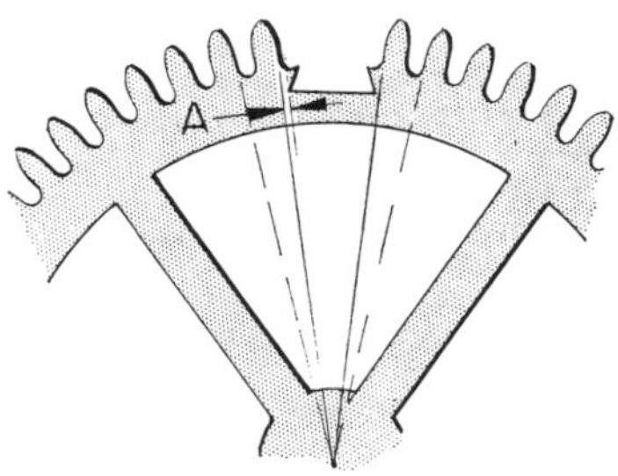

FIGURE 5
The depth of the dovetail should be about half the width of the wheel's rim. It's corner should not intrude on the adjacent tooth's radial line, A.

MAKING THE PLUG

The plug should be of hard sheet brass, slightly thicker than the wheel. Old clock wheels make good scrap from which to make these plugs. Place a large piece under the dovetail slot and secure it with a small toolmaker's clamp. Scribe the shape of the dovetail, as illustrated in Figure 6. Remember that the scribed line is slightly smaller and "inside" the actual shape of the dovetail slot. Therefore, cut the plug outside the line so it will fit snugly.

File the plug until it fits so tightly that no light can be seen through any crevices. Its protruding section should extend a little bit beyond the imaginary circle representing the full diameter of the wheel, as in A, Figure 7. Furthermore, the width of this protruding section should allow new teeth to be filed with their full width and length.

FIGURE 6
Clamp a piece of brass to the wheel and scribe on it the shape of the dovetail.

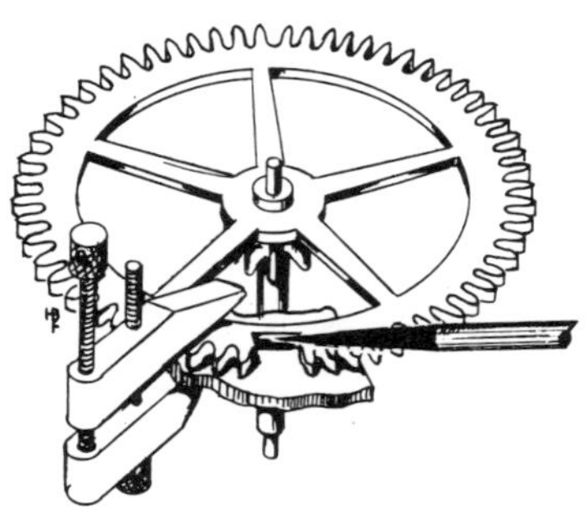

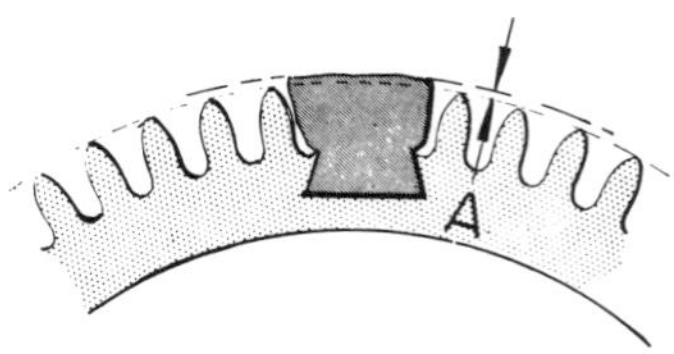

FIGURE 7
The plug should extend just slightly beyond the arc of the wheel teeth

INSERTING THE NEW PIECE

Securing the plug is done in two steps. The first is a light swaging of the plug. Remember that the sheet brass making the plug, *A*, is supposed to be slightly thicker than the wheel itself, *B*, as shown in Figure 8. Rest the plug on a solid, smooth, staking block, and lightly tap it a few times with a flat-faced hammer. This will cause it to spread out just enough to make it secure in the slot.

As a last assurance that the plug and wheel are as one piece, clean the wheel thoroughly. Smear the outline of the plug joint with soldering solution or a thin touch of soldering paste. Then place a small shaving of low melting soft-solder on the center of the crevice, as illustrated in Figure 9.

Hold the wheel high over the flame of an alcohol lamp so that the plug and surrounding area heat slowly. See Figure 9. As soon as the chip of solf-solder flows, remove the wheel from the flame so as not to soften the brass unduly. The solder will have flowed into all the reaches of the crevice by capillary action, solidly securing the plug to the wheel. Now file the plug to conform to the full circumference of the wheel.

FIGURE 8
The sheet of brass making the plug should be slightly thicker than the wheel itself. A few taps with a hammer will spread the plug to fit snugly.

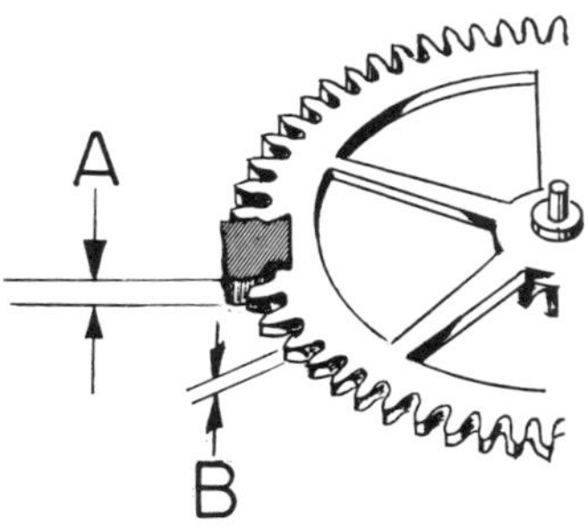

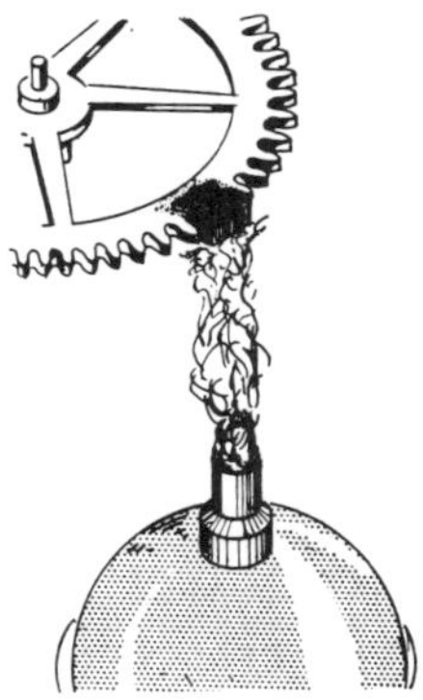

FIGURE 9
Smear the joint with soldering paste and place a thin chip of solder on the center of the crevice. Heat over a burner until the solder melts and seeps into the joint.

CUTTING NEW TEETH

Remember that you are not actually cutting new teeth but rather *spaces between* the teeth. In order to make certain of precise spacing, a further precaution is needed.

With the same type of brass from which the plug was made, choose a piece that will span about six teeth. Place this over a section of the wheel which has six good teeth and clamp it to the wheel. File this piece to the approximate outermost curve of the wheel (above the teeth). Next, using the six tooth spaces as guides, cut radial slots to the depth of the tooth slots, taking care not to cut into the wheel itself. This is shown in Figure 10. Scribe the curve of the inside rim of the wheel onto the adjacent piece as a further guide.

Now shift this guide over the dovetailed soldered plug so that an equal number of indexed spaces are on either side of the dovetail. The indexing saw cuts must match the existing tooth spaces perfectly, as in Figure 11. Clamp this accurately to the wheel, making certain that the scribed arc matches the rim of the wheel.

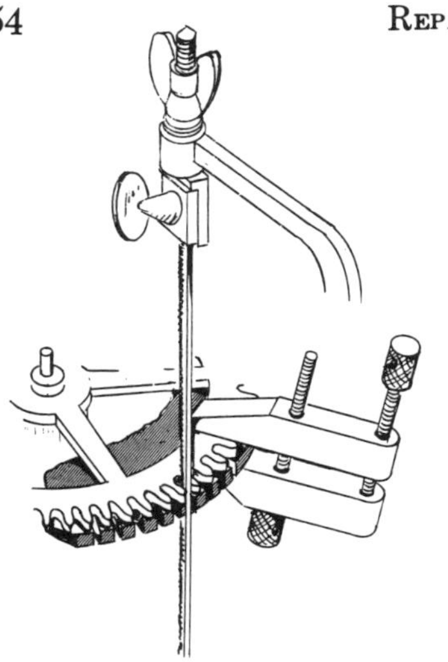

FIGURE 10
Clamping another piece of brass to a good section of the wheel, file the outer edge to an arc even with the wheel rim. Now cut radial slots the depth of the tooth slots. This will be your guide from which to cut the spacing between the new teeth in the plug.

Using this guide, place the saw blade in the cuts of the index-section of the guide and cut into the dovetailed plug. Your slots will be perfectly spaced, as in Figure 12. The width of the saw cut should approximate but not encroach on the full width of a tooth space. This is best left to be finished with escapement files.

The curves of the tooth tips also should be finished with escapement files, preferably those with safe edges so that the backs or sides of the files do not abrade or cut or mis-shape adjacent teeth. Filing the teeth shortens them. The reason for making the plug of thicker material now becomes apparent. The teeth can be stretched, if need be, by hammering a flat stake on top of the plug. Then they may be refiled and dressed to the correct shape.

FIGURE 11
Match the spaces of your guide perfectly with the spaces between the teeth on either side of the plug.

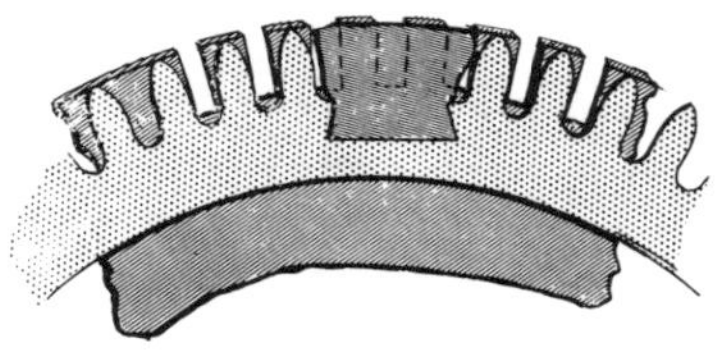

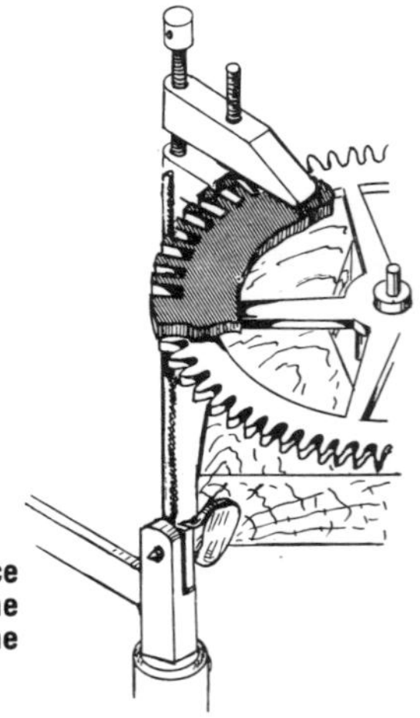

FIGURE 12
With the guide clamped firmly to the wheel, place the saw in the radial slots and cut into the plug. The new slots should be perfectly spaced around the wheel.

In Figure 13, two types of files are pictured. A triangular file shapes the curve of the replacement teeth. On this file, only the flat, bottom surface has a filing edge; the top two sides are smooth so as not to cut into the adjacent teeth. A small, round, needle file supplies the rounded bottom to the space between teeth. If the spaces are flat or square bottomed, slotting files with safe sides should be used.

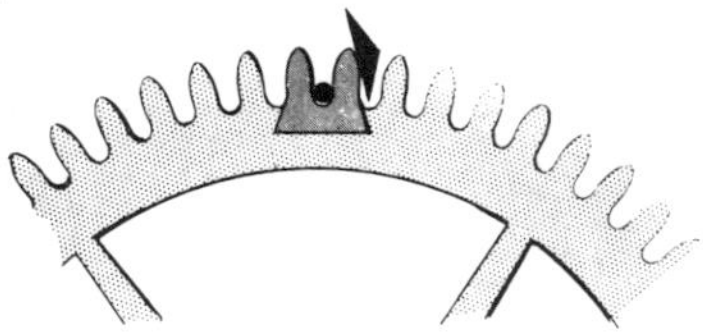

FIGURE 13
A round needle file shapes the curve at the base of the slot, while a triangular file shapes the tips of the teeth.

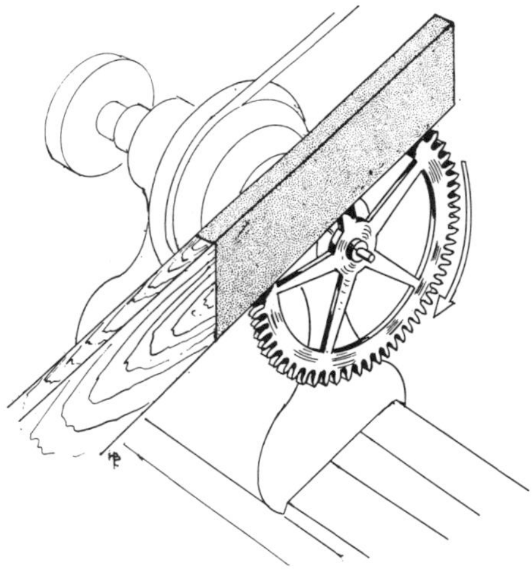

FIGURE 14
File the plug to the same level and thickness as the wheel. Then mount the wheel in a lathe and dress it with an emery stick.

To finish the job, file the plug to the same level and thickness as the wheel. Then mount the wheel in a lathe and dress it with an emery stick, as in Figure 14. Rotate the lathe in the opposite direction from that normally used to prevent your fingers from being injured by the rotating wheel teeth.

With the proper care, the results should make it difficult to tell where the repair was made.

INDEX

AMERICAN CLOCKS

A Guide to Identification and Prices

Volume 1

Tran Duy Ly

While there are many miscellaneous clock catalogues available, for the most part, these show products of a particular manufacturer at some specific period in that maker's productive history. Others, which include a greater selection of clocks encompassing most of the clocks made during the productive years of that factory, specialize in just one brand name. This new volume, in its 320 pages, is a handy encyclopedic reference which includes the products of nineteen manufacturers. The volume commences with an interesting history of the Ansonia Clock Co., written by Chris Bailey, a foremost authority on American clock history and former curator of the American Clock Museum at Bristol. He also supplies backgrounds for each of the other eighteen factories, these following in alphabetical order.

Thus, in one book are shown thousands of different clocks in various styles, movements, novelty cases and appearance. The experience of the editor, Tran Duy Ly, is evident in the expert development of the format in which each company's clocks are displayed. Furthermore, he has arranged the volume in a logical sequence so that random searching is not necessary, and finding a particular clock is made quite simple.

All clocks of similar type and appearance of each manufacturer are grouped together. In most other reproduced catalogues, a current value is printed adjacent to the pictured item. Immediately, this ages the volume as a price reference and can present misleading values in this era of inflation and ascending prices. In this book, each item has the date of the original catalogue in which each clock's illustration originally appeared. Prices are listed in a separate price list compiled by a national panel of experts and will be updated yearly.

The practical advantages and versatility of this volume on American clockmakers serve many purposes. First, it is a quick and ready reference of all manufacturers. Second, with its detailed index and cross references, it facilitates identification of specific clocks. Further, it provides facsimile reproductions from original factory catalogue art and not from old, faded photographs. Factory art confers a genuine reference of documentary value, a reliable reference with which to determine the originality of the offered clock and the ability to determine whether the timepiece has suffered alterations. Illustrations are sharp and clear, with precisely defined details, so much desired and sought after by expert restorers.

Mr. Ly notes also that catalogues from small companies such as Parker & Whipple, Warren Telechron, Russell & Jones, New York Standard Watch Company, and Tiffany Electric are generally printed in pamphlet form and easily become frayed and worn, misplaced or lost. This volume however, is presented in a securely bound edition and will be readily available and useful for many years.

Mr. Ly has also included an invaluable section addressing "IMPORTANT TIPS" which contains expert and useful advice and instructions regarding buying, selling investing, collecting, speculation and auctions. The novice collector as well as the more experienced clockmaker, dealer, restorer, and research historian will find this a valuable addition to the clock connoisseur's library.

HENRY B. FRIED

320 pages - 8 ½ x 11 - Smyth sewn Hardbound **$39.50**

CALENDAR CLOCKS

Tran Duy Ly

As a loyal and dedicated follower of American Calendar Clocks, what can be more fascinating than the mysterious movement of pointers and rollers at the appointed time and the resulting display of days, dates and months. How utterly amazing is this 100 year old mechanism doing its job in an accurate manner each day. How many modern 5 year old conveniences still work, let alone after 100 years.

Just how this minor miracle is daily accomplished is explained in clinical detail using ample photographs and drawings of cases, movements, wheels, rollers, hands, dials, labels and how to do it instructions. These heavily detailed drawings can be used for intricate repairs and the fabrication and placement of missing parts.

This 360 page volume covers 51 manufacturers and inventors listed including: Ansonia, Atkins, Boardman, Elias Burwell, Carter, Franklin & Morse, Feishtinger, Gale, Gilbert, Hubbell, Ingraham, Ithaca, Jerome and New Haven, Kroeber, B. B. Lewis, Macomb, Maranville, Parker, Prentiss, Seem, Sessions, Seth Thomas, Terry, Waterbury and Welch.

There are 41 pages devoted to the Ithaca Calendar Clock Company including a history of the company, directions for starting Ithaca spring clocks, directions for starting Ithaca weight clocks, and directions for correcting the Ithaca calendar if it should get displaced. This valuable information will be useful for many repairs and adjustments. Detailed photographs and drawings cover all aspects of Ithaca Calendar Clocks from the simple Farmer model to the rare Skeleton. There are detailed presentations on the 3 ½ Parlor, the Melrose, and several other Ithaca models. Rare custom design cases and the Cornell floor calendar clock are shown for the first time in any book.

Tran Duy Ly has made every effort to make this book serve as an accurate reference source for beginning through advanced collectors. He along with a panel of 54 leading national experts have carefully inspected every aspect of this text to insure the highest level of accuracy in pricing and authenticity of the clocks pictured in this book. Background information has been provided by Chris Bailey, a noted authority on American clock history and former curator of the American Clock and Watch Museum at Bristol, Connecticut, and James Baxter of Beech Grove, Indiana who has one of the most complete and exquisite calendar clock collections in America. It would be a thrill for all of us to be able to view his collection in person. Jim feels that this book is a ***"gotta have"*** for anyone who enjoys collecting and restoring clocks.

There is also detailed information on important topics such as dials, labels, hands, reverse glass paintings, patent drawings and pendulums.

If one is contemplating the purchase of a calendar clock, you will find all the detail needed to establish accuracy and originality as well as pricing guidance. There will be periodic updates so that this book will be a lasting source of reference. This new Calendar Book is another one of Tran Duy Ly's Reference Clock Books.

LARRY RAPPAPORT
Great Neck, New York

360 pages - 8 ½ x 11 - Smyth Sewn Hardbound **$39.50**

GUSTAV BECKER CLOCKS

A Guide to Identification and Prices

Tran Duy Ly

My very first antique wall clock was a Gustav Becker two-weight regulator, circa 1890. At the time, in 1969, I felt very ignorant concerning clocks and clock values. How I wish I had an excellent reference book like this one, *Gustav Becker* by Tran Duy Ly, to guide me in my purchase!

The Gustav Becker crown and anchor trademark is now quite familiar to clock enthusiasts. Usually shown on the dial, and virtually always on the movement, this mark assures a potential owner that the clock or timepiece was well-made, with strong attention to quality. It seems to me that Tran Duy Ly has successfully depicted such high craftsmanship in this new book, which sets its own mark for quality.

Gustav Becker (1819-1885) was a dedicated and successful German clockmaker. In 1850, thoroughly trained in clockmaking in Germany and in Vienna, he established a small clockmaking shop in Freiburg, Silesia (not the Black Forest Freiburg.) He struggled to produce clocks with an unskilled labor force until a design award in 1852 gave him the prestige to attract the quality of workers needed for success in this new industry. The firm prospered, and won many quality and performance awards worldwide from the 1860's to the 1890's. Until about 1880, virtually all Gustav Becker clocks were weight-powered regulator wall clocks similar to the one shown on this book's cover. Then, spring-powered clocks were introduced and many new styles followed. Almost four hundred models are pictured in the text. Several million clocks were produced and sold by the end of the trademark's use in the mid 1930's. Gustav Becker died in 1885, but the strength of his enterprise carried his name forward. The trademark and production survived an 1889 consolidation of Freiburg clockmaking companies into the "United Freiburg Clock Manufacturing Company Inc., formerly Gustav Becker". In 1926, this firm combined with the Junghans clockmaking company, and the Gustav Becker trademark continued until about 1935.

This volume demonstrates the great variety of clocks (and a few watches) manufactured by Gustav Becker. The clear pictures and current value estimates will be useful to potential purchasers or appraisers of Gustav Becker and similar German clocks. Equally important, the many views of movements, dials, hands, weights, gongs, and pendulums will help determine authenticity. The text is organized alphabetically by clock type, beginning with Alarm clocks. Regulators, both weight and spring-powered, are to be found in the Hanging Clocks section.

With this book, publisher and editor Tran Duy Ly continues his excellent program to provide educational and reliable clock reference and price guides. He, with the support of forty-two other experts nationwide, has made every effort to produce an accurate, informative reference for anyone interested in Gustav Becker clocks.

Whether you are just beginning to collect, wish to know more about a clock which you already own, or simply enjoy having a good clock reference library, you need this book. Mr. Ly will also publish periodic up-dates of prices for every clock in this book so that this guide will never be outdated. Thus, this *Gustav Becker* book becomes an even better choice with time and allows the clock connoisseur to make more intelligent decisions when it comes to buying or selling.

DOUG COWAN
Cincinnati, Ohio

8½ x **11** - **248** pages - Perfectbound **$29.50**
Smyth sewn Hardbound on white gloss enamel text **$35.00**

LONGCASE CLOCKS AND STANDING REGULATORS

PART 1: MACHINE MADE CLOCKS

Tran Duy Ly

Every clock collector's dream is to possess one magnificent clock, unequaled, the culmination of his or her desires. For many, that dream clock would be a long case clock or standing regulator proudly displayed in the entrance to their home. With this excellent guide, Tran Duy Ly will prepare you in the quest for your dream clock.

Long Case Clocks and Standing Regulators takes the serious collector and investor through hundreds of clocks made by no less than fifty precision companies from several different countries. There is no other work available that provides nearly the coverage of this excellent reference.

The volume begins with a well written and complete list of important buying tips by Thomas Spittler, a recognized authority on European and American long case clocks. He covers important questions and considerations that one should familiarize themselves with before purchasing one of these clocks.

This book not only covers the general description of the various pieces but also provides the technical information necessary to allow the collector to be thoroughly informed prior to making a purchase. More importantly, the subtle differences that make some clocks more desirable and thus more valuable are well presented.

It is quite unlikely that any of us has heard of Haas & Shone, Harris & Harrington, Hirst Brothers & Company or many of the other companies listed in this very large and complete volume. On the other hand, clocks by the well known companies such as E. Howard, Seth Thomas and Colonial are covered in complete detail. Many of their models are in no other reference.

In reading through this book, one realizes that Mr. Ly has composed the most complete and comprehensive reference on these clocks. It is evident that he has spent hundreds of hours researching collections and horological literature to obtain the massive amount of information necessary for this very complete work.

Serious collectors, dealers and restorers are fortunate that Tran Duy Ly has taken the lead in horological reference books and price guide information. As usual Mr. Ly has gone to the best informed and experienced collectors and dealers to obtain realistic values for these clocks.

While going through the 504 pages one soon realizes that the machine made, factory produced long case clocks and regulators were made in vast numbers both in America and Europe, and yet when one is found either in the back room of some old antique shop or offered in some auction house this book will surely guide you through the mystery of these wonderful clocks.

DAVID MORGAN
Star Fellow NAWCC
President NAWCC

504 pages - 8 ½ x 11 - Over 1150 black & white illustrations.
Smyth sewn hardbound in high gloss enamel text............**$69.50**

NEW HAVEN CLOCKS & WATCHES

With a Special Section on New Haven Movements.

Tran Duy Ly

I began to seriously collect clocks while living in England in 1974 and most of the clocks I found were American clocks by the New Haven Clock Company and its associated British sales firm Jerome & Co. Many of these clocks had an interesting label showing a factory identified as "The New Haven Clock Company formerly the Jerome Manufacturing Company." I wish I had had a book such as ***New Haven Clocks & Watches*** by Tran Duy Ly to help me better understand the history of the New Haven Clock Company and identify the many clocks and watches they produced.

Tran Duy Ly has produced ***New Haven Clocks & Watches*** to complement his other books on Seth Thomas clocks, Waterbury clocks, Ansonia clocks and Welch clocks. In my opinion, these five firms combined to produce over 75% of the clocks manufactured in America. This exciting new volume on New Haven clocks is produced to the usual high standards of the Arlington Book Company. It contains an amazing 38 chapters and over 500 pages on different types of New Haven clocks and a supplementary price guide. The many views provided will help identify correct dials, hands, pendulums, weights, gongs, movements and trim.

What intrigues me most about this book is the 11 page history and chronology written by the noted horologist Chris Bailey. Chris clearly develops the story of the giant of American clockmaking, Chauncey Jerome, and his relationship with Hiram Camp, the first president of the New Haven Clock Company. I have always felt that if Camp had named the firm the Chauncey Jerome Clock Company he would have captured its true identity in the way the Seth Thomas Clock Company did. But this did not happen. This caused problems for the fledgling New Haven Clock Company when it tried to sell its products in England. The British identified the name "Jerome" with quality American clocks and were willing to accept no substitutes. To overcome the situation New Haven used the name of their independent British sales firm, Jerome & Co., and New Haven clocks with the Jerome & Co. label were successfully sold in England from 1856 to 1904. Even today the name "Jerome" in England is bigger than Seth Thomas and E. N. Welch combined.

So it is fitting that the Arlington Book Company has filled a void by producing this long awaited book on the New Haven Clock Company. Tran Duy Ly, more than any other author of horological books in America, has repeatedly produced books that are right on target for clock collectors around the world. Each new book builds on the solid foundation he has established in the past. ***New Haven Clocks & Watches***, like his other books, is well written and easy to use. It contains over 2,200 quality illustrations and photographs. The section on kitchen clocks contains the most models ever seen in print. In addition to hundreds of pages of clocks there are even 24 pages of New Haven watches.

I feel the 50 pages of movements and parts showing extremely clear and detailed illustrations of 61 New Haven movements are outstanding. Most of these pages show a front and rear view of the movements and for the first time I can recall, each movement part is clearly identified. For the repairman, these 50 pages alone are worth the price of the book.

In summary, I am truly excited about Tran Duy Ly's new book ***New Haven Clocks & Watches***. If you own a New Haven clock and wish to find out more about it, or if you deal in clocks and need to know up-to-date prices, or if you repair clocks and could use wonderful illustrations of

TOM SPITTLER
New Carlisle, Ohio

8 ½ x 11 - **520** pages - Perfectbound **$47.50**
Smyth sewn Hardbound on white gloss enamel text **$55.00.**

SETH THOMAS CLOCKS & MOVEMENTS

Tran Duy Ly

Ten years ago Tran Duy Ly published his first edition of *Seth Thomas Clocks & Movements* to stunning reviews. The book was greatly commended for many reasons, none more important than it being the first real comprehensive guide to the identification and values of most Seth Thomas clocks and movements manufactured during the firm's first century of production since 1810. The pages of his *Seth Thomas Clocks & Movements* contained refreshing, exact reproductions of Seth Thomas clocks found in original catalogs. Also, the arrangement of clocks was logical and indexed for ease of use — reflecting the publisher-editor's own insight as a highly knowledgeable and respected collector.

Thus, on hearing of the release of his second edition of *Seth Thomas Clocks & Movements* my expectations were high to say the least. Mr. Ly has also added much more new material and meticulous detailing to the vast collection of clocks already there. Highly precise and clear photographs have been added to the ensemble of catalog reproductions and many illustrations have been enlarged. As is the case with all his publications, there is a supplementary price guide which includes current, realistic values of each clock as determined by a national panel of the best-informed and experienced experts on Seth Thomas clocks. The editor-publisher has thoughtfully separated the price from the book thereby allowing their updating — as values of clocks are never static.

Chris H. Bailey, a prominent authority of American clock history and former curator of the American Clock Museum in Bristol, Connecticut, has contributed an elegant 10-page history of Seth Thomas and his clock company. There is an eye-appealing City Series section which contains clear and crisp illustrations of models not included in the early edition such as the Alton, Cairo, Erie, Lafayette, Peoria, and Topeka. A complete treatment of Calendar clocks is also provided including Fashion calendar clocks. This publication is also enriched by the knowledge of additional experts such as, Thomas Chapell on the topic of Sonora Chime clocks, Charles May on Fashion Calendar clocks, Tom Spittler on Longcase clocks and Mantel Clocks-Wood Cases and Paul V. Heffner on the dating of Seth Thomas clocks. Mr. Ly has again included a separate, thorough and more vast Movements section. Hundreds of illustrations and descriptions of Seth Thomas movements, along with a cross-referenced table, are also provided.

Mr. Tran Duy Ly's newest book, the second edition of *Seth Thomas Clocks & Movements*, holds true with the precedence of excellence set forth by his many previous publications — those who own one are sure to agree with me. In recent years, Seth Thomas clocks have become evermore rare and collectable and this new guide is sure to serve as the perfect collecting tool for us all.

DAVID WARNER, FNAWCC

2nd Edition Revised and Enlarged in 1996

496 pages - 8 ½ x 11 - Paperback **$39.50** - Smyth Sewn Hardbound **$49.50**

CLOCK WHEEL & PINION CUTTING

by **J. Malcolm Wild**

This little book comprehensively covers how to calculate and cut wheels and pinions on a conventional lathe and is written by an expert in professional clock repair.

48 pages - 5 ½ x 8 ½ - Perfectbound **$11.95**

WHEELS & PINIONS

& How to Determine Their Exact Size

by **H. Shouffelberger & T. Gribi**

This book contains nine tables and one plate which are all that is necessary for the watchmaker to determine the exact size of missing wheels or pinions. The tables apply to both systems of measurement i.e. the English inch or metric system and are more convenient and faster to use than the old method.

20 pages - 8 ½ x 11 .... **$5.00**

PRACTICAL BALANCE AND

HAIR-SPRING WORK

BY WALTER J. KLEINLEIN

Great instruction and demonstrations to increase the workers understanding in handling the balance and hair spring associated with fine watches. Highlights practical features that are associated with the daily work of the repairer including getting a balance true and in poise or fitting a new hair spiring, reforming and overcoil or truing a damaged spring. An absolute bible for any watch repairer!

115 pages - 5 ½ x 8 ½ **$11.95**

Manual for Overhaul, Repair & Handling of **HAMILTON SHIP CHRONOMETER** with Parts Catalogs.

Additional of May 1st, 1946 Repair Parts List by **US Navy Department**

The Hamilton Model 21, 14-jewel ship's chronometer was standard issue for all U.S. vessels for many years. This reprint of the Navy Department's manual for disassembly, repair and reassembly contains easily followed, well illustrated step-by-step instructions for this important instrument.

104 pages - 8½ x 11 - Perfectbound **$14.95**

Manual for Overhaul, Repair & Handling of US Navy Mechanical, Boat & Deck Clocks

CHELSEA TYPE with Parts Catalogs. by **US Navy Department**

This superbly illustrated manual on the repair and handling of Chelsea Boat and Deck clocks covers time-only Model 12E and 17E. Nonetheless, much of the material is equally applicable to the time train of many Chelsea striking models.

10 pages - 8 ½ x 11 - Perfectbound **$14.95**

Manual For Overhaul, Repair and Handling of

HAMILTON 35-SIZE

Chronometer Watch

with Parts Catalog. - Navy Department

This subject covered include: Introduction; Description; Preparation for Use; Principles of Operation; Disassembly of Chronometer Watch; Cleaning; Inspection and Repair; Reassembly of Chronometer Watch; Test and Adjustment; Storage and Shipment; Maintenance Parts Catalog; Numerical Parts List and Service Tools.

68 pages - 8 ½ x 11 .... **$9.95**

The Watchmaker's & Model Engineer's Lathe

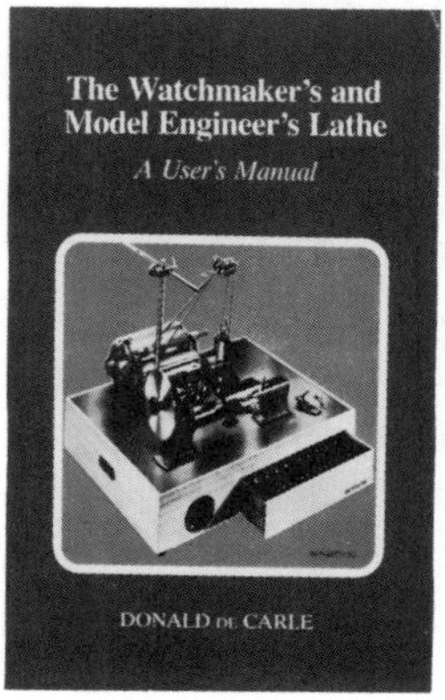

A User's Manual by **De Carle, D.**
Revised by **Eric Restall.**

This book has become a standard work on the subject. The bulk of the text deals with the basic construction, selection, care, maintenance and accessories for the small precision lathe and its uses.
This new edition is unique since it brings together a wealth of information on small precision lathes, both past and present.

195 pages - 5 ½ x 9 - Hardbound **$29.95**

STAKING TOOLS

AND HOW TO USE THEM

By K & D Manufacturing Corp.

Contains a brief history of the development and use of the Staking Tools set. It also contains the first several pages of the 1910 Catalog of Staking Tools and constitutes a hand-book every clockmaker needs on his bench on the various jobs related to the use of the Staking Tool.

108 pages - 5 ½ x 8 ½ - Perfectbound **$11.95**

Watch Repairing Cleaning and Adjusting

by **F. J. Garrard**

A Practical Handbook dealing with the Materials and Tools Used. Turning, Making Small Tools, Barrels, Fusees, Main-springs & Chains, Depths, Train Wheels, Escapements, Balances and Hairsprings, Adjusting & Timing, Motion Work, Hands & Dials, Cases, Keyless work, Causes of Stoppage of Watches, Conversions and Alterations in Movements and the Methods of Repairing, Cleaning all kinds of Watches, Repeaters, Chronographs and Marine Chronometers.

214 pages - 5 ½ x 8 ½ - Perfectbound **$16.95**